The Teaching of Vedic Mathematics

Dr. S.K. Kapoor

Lotus PRESS
4735/22, Prakash Deep Building
Ansari Road, Darya Ganj,
New Delhi-110002

Lotus Press : Publishers & Distributors
Unit No. 220, 2nd Floor, 4735/22, Prakash Deep Building,
Ansari Road, Darya Ganj, New Delhi- 110002
Ph.: 32903912, 23280047, 9811838000
E.mail : lotus_press@sify.com

The Teaching of Vedic Mathematics

ISBN: 81-8382-056-5

Published by : **Lotus Press Publishers & Distributors,** New Delhi-110002
Printed at : Afcons Enterprises, New Delhi

INNER VOICE FOREWORD

— FROM A SAINT

Hear your inner voice and be free from the outside limitations. Within is the seat of the Lord and it is here where we have to be face to face with our selves. This origin source reservoir is to be tapped, and this appears to be the urge and intensity of present attempt to reach at 'Vedic mathematics basics' in five steps. These five steps would prepare for the phase and stage to be in communication with the SELF.

The SELF is TRANSCENDENTAL. The seat of the SELF within *Triloki* is the Creator's space. Self established at the seat of Creator's space ensures carrying the Being along the transcendental carriers. It is the bliss of this ensured supply of transcendental carriers which intensifies the urge of the transcending mind to glimpse the inner folds of the transcendental worlds manifesting within Creator's space.

The whole efforts and beginning for it, as such is to be to attain transition from Triloki manifesting as outside mundane macro state world as first state of consciousness for the intelligence to the next transcendental micro state world of the intelligence lively within Creator's space as origin source reservoir of Triloki. Our ghan / body cube, as such, is to be chased for its origin source reservoir within it lively as Chatar mukhi / four head Creator.

Really blissful is the learning path of this chase range. It is transcendentally blessed. Let sadkhas permit their transcending mind to fulfill with bliss of the transcendental worlds / Panch Mukhi / five head lord sustaining the Creator's space.

Lively within
cavity of heart:
25-11-2005

In search of ***JYOTI***
manifesting as **NAD**

■

PREFACE

Urge to know ancient wisdom expressed as Vedic knowledge inherently imples acceptance for chase of it the way these systems permit. This as such would mean that the sadkhas for the chase of Vedic systems have to proceed along the established paths of *Sankhiya Nishshta* and *Yoga nishta* which not only run parallel to each but are also complementary and supplementary of each other.

The sankhiya nishtha, on its ultimate analysis amounts to availing the artifices of numbers, for coordination arrangement of organisation format, while on the other hand, the Yoga nishta avails the dimensional formats, for regulation of transcendence and ascendance of mind through the manifested creations. The **TRANS**, also means, the number ahead, and it is this way, the transcending mind ultimately avails the artifices of numbers to sequentially glimpse the ultimate core of the transcendental worlds.

The par-excellence feature of Vedic systems is that these approach the whole range of knowledge as a single Discipline, and as such, on ultimate analysis the organisation of Discipline of knowledge avails the dimensional frames parallel to the artifices of numbers. This, this way, shall be taking us to the artifices of numbers, and the dimensional frames as the source processing tools in terms of which even the self referral feature of transcending mind chasing its own transcending path can also be worked out intellectually.

It is this feature which deserves to be chased for the basics, and here in the present attempt, the urge being to learn and teach Vedic mathematics on its basics, as such, at this phase and stage of exposing young minds during Middle school learning, The focus is to be upon the centre of the cube / origin of 3-space being of the format of hyper cube-4 and as such seat of creator's space.

This focus naturally aims to eye the goals of attainments for the young minds for perfection of their intelligence by having transition from the set up of the Cube to transcend ahead for glimpsing the set up of hyper cube-4 along with its transcendental base of the format of hyper cube-5.

Though, the learning all about hyper cubes 4 and 5 is to be the content focus of high school classes, and ahead there to would be the subject focus of the formats of hyper cubes-6 and 7 during higher secondary classes, but the initiations for the foundations of appropriate mental state for those formats are to be ensured during schooling of middle school stage.

This way the learning range of this phase and stage of schooling during classes 5 to 8 is to be centred around the set up of the cube with focus upon its centre /

origin of 3-space as seat of creator's space (4-space) as of format of hyper cube -4. Accordingly the teaching and learning of this range of schooling is phased as five books namely, book-1The teaching of Vedic mathematics, book-2 Learning Vedic mathematics on first principles, book-3 Vedic mathematics basics, book-4 Vedic mathematics skills and book-5 Vedic geometry course, for perfection of intelligence.

These books, in their integrated approach of chase of the range for set up of the cube / 3-space with focus upon its center / origin of 3-space, as source reservoir for the features of the set up, is meant for exposure of young minds to acquire perfection of intelligence while learning basis of basics on first principles along the concrete format of a cube.

The first principles for approaching basis of basics of Vedic mathematics on solid formats are being supplied by the Ganita Sutras, particularly as the ordering principle of Ganita Sutra-1 which makes it as the source Sutra, and as such 'cube' and 'Ganita Sutra-1' are to ever remain within the eye sight of the learners of this phase and stage of schooling.

The aim and goal of learning being the chase of basis of basics on first principles to ensure perfection of intelligence for the young minds, as such heavy responsibiliy lies upon the Vedic mathematics teachers and parents to make the learning really blissful throughout the duration of chase.

The whole range of enlighened souls begining with Maharishi Valmiki and Maharishi Ved Vyas to Maharishi Dayanand and Swami Bharti Krsna TirthaJi Maharaj to Shri Pad Baba and Maharishi Mahesh Yogi, this learning path is being made blissfully lively, and the enlightened teachers are to continue the linege.

With these inspired background, present humble attempt has been made to reach at these books and it is hoped that these may provide required exposure for the young minds with the help of the enlightened teachers, who shall be making their instructions free of technicalities and the knowlede content to flow blissfully within the head s and hearts of the students.

Mr. Kenneth R. Williams and Mr. R. P. Cosmic Kapoor deserve my heartful thanks for remaining intellectually with me during these days of my reaching at these books. Mr. Rohtash Kapoor deserves full appreciation for conducting Vedic geometry course.

Mr. Deepak Girdhar deserves all appreciation for his nice graphics and the computer work.

Credit for nice publication goes to Lotus Press.

The inner voice forward from a saint gives me the satisfaction and confidence as that these efforts may be fruitful for the cause of basics to be brought home to the young minds.

Vedic home, **Author**
Rohini, Delhi
25-11-2005

■

CONTENTS

SECTION - I
ART OF TEACHING

1

TEACHING RESPONSIBILITY

1. The teaching responsibility of this phase and stage of the Discipline of Vedic mathematics is for perfection of intelligence of the students for comprehension and chase of the basis of basics of Vedic mathematics on their first principles.
2. This responsibility, as such, essentially amounts to bringing within the comprehension of the students as that the 'basics of Vedic mathematics' are nothing but the features of the set up of cube(⊞) as representative regular body of 3-space.
3. Further, this responsibility would amount to bringing home to the students as that the centre of the cube, as origin of 3-space and also as seat of Creator's space (4-space / ⌑) is the sole point source reservoir of the basics and as such is the basis core of the basics.
4. Still further the responsibility of teaching would extend to imparting instructions for working out the basis of basics on their first principles which are inherently embedded into the set up of the cube as its coordination arrangement of the

organization format permitting content flow within domain of the cube.

5. This, as such, takes to the learning skills for flow of content within domain of the cube, as is framed and enveloped within the synthetic set up of eight corner points, twelve edges and six surfaces.

Synthetic geometric envelope of cube of 26 entities

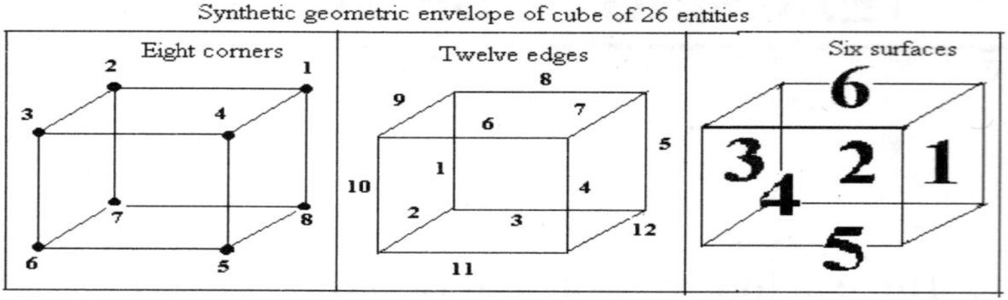

6. The proper discharge of teaching responsibility, in the circumstances, would mean to help the students to learn to comprehend this manifestation of the set up as 'cube' and to follow it for the skills for its applied values as well as for perfection of intelligence of the students.

7. As such, the focus of this responsibility, and accordingly of the whole exercise of 'teaching and learning of this phase' is to get focus upon the set up of the cube as it is.

8. 'The cube', being the in between, transitory position, with square (■) on its one end and hyper cube 4 (⌗) on its either end, therefore, the teaching feature should take note of it, so that the teaching and learning of the set up of the cube in no way comes in the way of attaining smooth transition from the set up of a cube to that of the set up of hyper cube 4.

9. The teaching and learning of this phase is to attain melting of the mental format of linear order thinking inherent in the set up of a cube to transit

to that of the spatial order thinking format embedded inherently in the set up of hyper cube 4.

10. The success of the learning of this phase and stage of schooling is not to be measured in terms of the comprehension and application of the features of the set up of a cube but in terms of the success of melting of the mental format of linear thinking for desired transition for acquisition of spatial thinking of process of the spatial order embedded into the set up of hyper cube 4.
11. It is this attainment of transition for the mental frame which shall be helping deciphered the distinguishing features of all the points of the domain of cube from the centre of the cube.
12. This, as such would emerge to be a self certifying evaluation for the students about the success of their learning in the degree of their comprehension of the distinguishing features of center of the cube/origin of 3-space being a seat of Creator's space/4-space from that of any other point of the 'domain of cube' being simply of the features of the set up of 3-space domain.
13. It is this comprehension of the centre of the cube / origin of 3-space as seat of 4-space being of spatial order while every other point of the cube is simply being of features of linear order only, which ensures the success of learning of this phase and stage of the schooling.
14. As such, the teaching, naturally is to be initiated with this aim the students attain the requisite transition for the linear order mental format to that of spatial order potentialities of the mind.

■

2

MENTAL STATE OF THE TEACHER

1. Though the teaching and learning are inherently intimately connected as both being the 'same subject content centred' but the expectations of the teaching responsibilities would take the mental platform / state of teacher to be of one step ahead of the attainment to be had from the instructions to be imparted by the teacher.
2. The expectations of the teaching responsibilities of this phase and stage being to ensure attainment for the students 'a perfection of intelligence in terms of which there would be required transition and transformation for their mental set up of a linear order thinking of 3-space set up to that of the spatial order thinking of 4-space set up.
3. As such, as the instructions to be imparted by the teachers are to be to ensure attainment of such transition for the linear order thinking mental format of students to that of spatial order thinking mental format for them, so naturally the mental format of the teachers themselves must be of the order which have not only already attained and

glimpsed the spatial order of 4-space but also have attained required transition and transformation for their spatial order format to attain the solid order format of the transcendental worlds (5-space / ௐ).

4. It is this expectation from the teachers which takes them to the transcendental format for making the students to be fully knowledgeable about the manifestation formats of creations, of which cube is first such manifested domain.
5. The transcendental features of the mental format of teachers shall be making them capable of chasing the transcendental phenomenon of emergence of solid quantifiers within Creator's space from its transcendental core.
6. This transcendental phenomenon is of the range of emergence of solid quantifiers and their manifestation within the Creator's space as of the format of hyper cube 5.
7. The solid quantifiers, as those emanate from the transcendental core of Creator's space, flow into the domain of hyper cube 4 because of transcendental base of Creator's space playing the role of origin fold of hyper cube 4.
8. The solid quantifiers fulfilling the 4-space domain are superimposed upon the spatial order of 4-space, like the spatial grids getting superimposed upon another spatial grids with the grid joints of the superimposed grid being the manifested knots at the centres of the grid zones of the base grid.
9. This glimpsing of the transcendental phenomenon by the teachers shall be helping them for making

the students to learn by chasing the set up of 3-space in its static position but the boundary surfaces may be having degree of motion which may help cross the domain in between the pair of surfaces by availing the arrangement of superimposition of spatial grids and the content flow stream through the centres of the base grids.

10. This glimpsing of the transcendental phenomenon, as such further shall be helping the teachers to discharge their responsibility of teaching this phase and stage of learning of Discipline of Vedic mathematics of basics as features of the cube as 3-space domain and skills for chase of basis of basics on first principles by following the coordination arrangement of flow of streams through the centres of the base grid zones as a format for the surfaces of the cube.

11. As such the mental platform of the teacher and expectations of responsibility of teaching takes the teacher to the phase and stage of transcendental attainments, as much as that they having attained transition and transformation for the spatial order of intelligence field to glimpse the solid order of the consciousness field emerging within the intelligence field within Creator's space.

12. This, this way establishes the mental state of the teachers within the consciousness field for ensuring attainment for the mental frames of the students for their establishment within the intelligence field.

13. To conclude, the expectations of the teaching responsibilities takes the mental state of the

teachers being established within the solid order consciousness field.

14. This expectation for the mental state of the teacher to be established within the solid order consciousness field means as that the teachers have already glimpsed the transcendental phenomenon of emergence of transcendental worlds () within Creator's space ().

15. The transcendental worlds are the solid order set ups and are designated as 5-space features domains.

16. As such the teachers being of transcendental state, they are fully capable of imparting instruction to the students for ensuring perfection of intelligence for them to have requisite transition and transformation for linear order thinking for glimpsing the spatial order thinking format. ■

Technical terms up to date

1. Creator's space is 4-space.
2. Transcendental worlds are 5-space.

Symbols up to date

1.  Cube / 3-space
2.  Hyper cube-4 / 4-space
3.  Hyper cube-5 / 5-space

■

3

FOUNDATIONS FOR THE MENTAL STATE

1. The foundation for the mental state of teachers, which is to be of transcendental features, is to be of the order of the set up of the transcendental core of the Creator's space (⛶).
2. The transcendental worlds being of solid order, as such the transcendental base of Creator's space is of features to sustain the spatial order of the Creator's space in terms of the solid order of the transcendental worlds (ऊ).
3. The transcendental base of the Creator's space being of solid order, the same fulfils the Creator's space with solid quantifiers supplied by the solid order of the transcendental worlds.
4. The solid quantifiers supplied by solid order of the transcendental base of Creator's space transcend and flow into the Creator's space domain through its origin / transcendental core as seat of the transcendental worlds as base of the Creator's space.
5. Therefore the foundation of mental state of the teachers is to run parallel to the set up of

sustenance of the Creator's space with transcendental base supplying solid quantifiers though the transcendental core into the Creator's space domain.

6. This transcendental feature is depicted and preserved in the scriptures, particularly in *Shiv Puran* as that this foundation is establishing *Shiv Lingam* into the mystic Creator's domain.
7. Symbolically, it may be depicted as follows, as a cube, as solid structure at centre of hyper cube 4

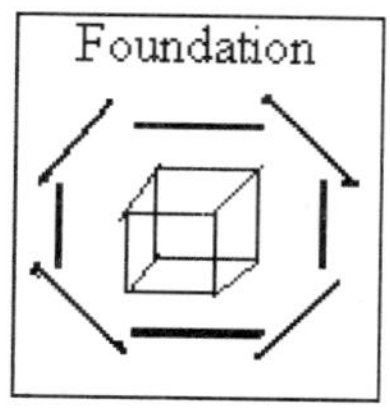

8. This as first step of the transcendental phenomenon and the manifestation of it as world of transcendental features within Creator's space, may be depicted as follows:

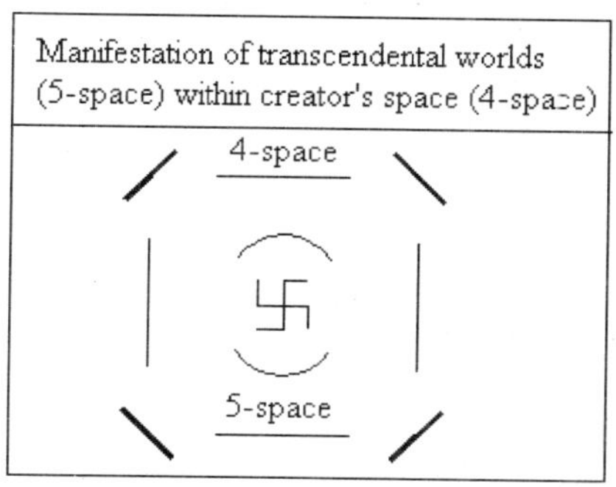

9. The transcendental features domain within Creator's space splits into a pair of hemispheres,

and this phase and stage, of transcendental phenomenon, may be depicted, as follows:

Northern hemispher

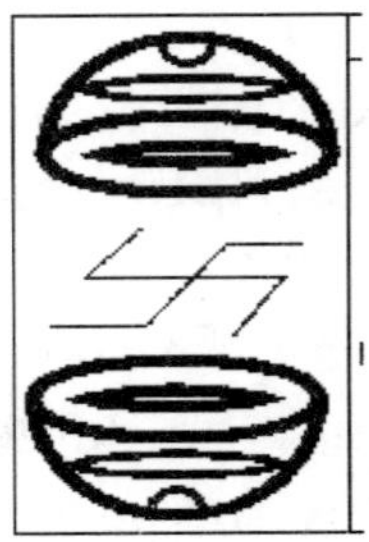

Southern hemisphere

10. As such, the foundation for the mental set up of the teachers, and the range of the transcendental phenomenon of this foundation, shall be making the attainments of the teachers to be of such order which shall be enriching their art of teaching for the science of learning of the students; learning, which is to begin with the static position of a cube.

■

Technical terms up to date

1. Transcendental core is origin of 4-space
2. Mystic / Creator's domain is hyper cube-4
3. 'Shiv Lingam' is format of hyper cube-5

■

4

CONTINUOUS GLIMPSING PROCESS

1. The mental state expectations from the teacher, not only from the point of view of teaching responsibility but also from the point of view of teacher's own bliss and enlightenment, there is a need of continuous glimpsing of the inner folds as well as inner most fold of the transcendental worlds.
2. The continuous glimpsing process will attain unity state of consciousness and with its attainment the transcending mind shall be riding the transcendental carriers which shall be leading through the core of the transcendental worlds to the selreferral domains of the order of the orb of the Sun.
3. The process and technique of transcendence deserves to be instructed to the *sadkhas* who shall by availing it would be attaining the requisite unfoldment of the set up of Creator's space / 4-space / hyper cube 4 and centre of the cube / origin of 3-space.
4. The continuous transcendence on the part of the

sadkhas shall be firstly attaining transition from the linear order of 3-space and secondly it shall be attaining the spatial order of 4-space.

5. The continuous transcendence on the part of teacher in the transcendental mental state, and with glimpsing of the inner folds and inner most fold of the transcendental worlds, shall be attaining the hyper solid order of the selfreferral domains parallel to the hyper solid order of orb of the Sun.
6. As such, the continuous glimpsing process of the sequentially compactified consciousness states in sequence and order of compactification, becomes the basic mental process for the required intellectual mental states.
7. It is the consciousness states, parallel to which only, are to be available the intellectual states for the *sadkhas*, and therefore, the intellectual attainments are to be expected only parallel to the order of the experiential gains through continuous glimpsing process of the transcending mind.
8. One can hope to transcend through the manifestation formats of creations only by the continuous glimpsing process of the transcending mind and this being so, before one undertakes the intellectual exercise for chasing the Reality lively ahead of the manifested domains, the experiential glimpse through transcendence is bound to be gained as a pre-requisite.
9. The intellectual chase of Reality as a set up of hyper cube 4 ahead of a set up of a cube, and that at the centre of the cube / origin of 3-space as

seat of Creator's space / 4-space / hyper cube 4 would be of fruitful comprehension only with continuous transcendence exercise of the mind. It is only through this continuous transcendence exercise that the existing mental frame can be melted for 'picks' transition and transformation for new mental frame in place of the old mental frame.

10. Therefore, the learning process of the young minds itself, which is part of art of teaching, is to be dually taken care of by the teachers undertaking responsibility for teaching Vedic mathematics basics at this phase and stage of schcoling for the young minds of the biological age group of 10 to 14 years of age.
11. This responsibility of imparting instructions as to be the process of transcending can be discharged only by personal supervision of the transcendence exercises of the *sadkhas*; at least during initial stages of the performance.
12. The teachers shall join the students during their 'sittings of trans', may those be at individual level or at collective level.
13. The sharing of the 'trans-experiences' with a teacher is one privilege which should never be permitted to go unveiled, as it is this sharing of experiences which may help the teacher as a senior *sadkha* to lead for further trans sittings.
14. The other way, which would be left with the *sadkhas* (including teachers), would be that one shall have selfevaluation of one's trans experience

in the light of the experiences of senior *sadkhas* preserved in the available Scriptures.

15. The other situation may be in the form of satisfaction in the light of consequential intellectual attainments of the individual, as intellectual attainments are to run parallel to the glimpsing attainment of the transcending mind.
16. However, there may or may not be an opportunity for evaluation of transcendence experiences, the bliss which one shall be gaining during 'trans-sittings' may be taken as the utmost evaluation certification of the *sadkha* being on the right path.
17. The duration for which and the frequencies at which one shall be remaining in blissful trans sittings, in itself is the certification in itself. ■

Technical terms up to date

1. 'Trans' also means 'number ahead'
2. '*Sadkhas*' also means students
3. 'Picks' are like 'solid whole entities', and also these are 'as fractions of flow paths'.
4. Intellectual interaction of mind is while it is 'not in trans' while experiential interactions of mind are while it is 'in trans'.
5. Transcending mind means minds state during 'trans'

■

5

PARALLEL INTELLECTUAL ATTAINMENTS

1. The experiences of the senior *sadkhas* preserved in the scriptures take us to the sequential compactification of seven states of consciousness and parallel to them being the intellectually very fine attainments as set ups of the geometric formats of these consciousness states.
2. The seven fold states of consciousness are designated as (1) waking state (2) dream state (3) deep sleep state (4) turia state / transcendental state (5) cosmic state (6) God state and (7) unity state.
3. The features of the geometric formats of expressions of these states of consciousness are of the nature of linear order attaining waking state world (3-space), the spatial order attaining dream state world (4-space), the solid order attaining deep sleep state world (5-space), and the hyper solid orders attaining subsequent states of consciousness in the sequence and order of their compactification range within the consciousness field.

4. Beginning like that with 3-space as waking state world, the range of consciousness states lead up till 9-space as the ultimate Brahman domain.
5. In this organization, the beginning with the waking state of consciousness, as linear order – 3-space, shall be bringing within the intellectual comprehension as such as a linear order set up of 3-space with cube as its representative regular body.
6. As such, the *sadkhas* permitting their minds to transcend shall be melting the linear order of their mental block and shall be transiting from it and would be transforming for attaining the spatial order (■) as a mental block.
7. The intellectual chase of spatial order 4-space would be facilitated for its comprehension for those who shall be through the continuous transcending exercise for glimpsing the inner folds and inner most fold of the Creator's space.
8. The experiences of senior *sadkhas* of glimpsing of the inner folds and inner most fold of Creator's space are preserved in the scriptures as that the set up of hyper cube 4 is of the features of the format of idol of Lord Brahma, four head lord, with a pair of eyes in each head, Creator the supreme, sitting majestically upon a lotus seat of eight petals. Lord Brahma sitting majestically on its lotus seat meditates gracefully within cavity of its heart upon its Lord, Lord Shiv, five Head Lord, with three eyes in each of the heads, the Lord of transcendental worlds.
9. This parallel experiential and intellectual chase

expression preserved in the scriptures as the attainments of senior *sadkhas* brings to focus the features of the set up of hyper cube 4 as a spatial order 4-space domain of transcendental core being enveloped within solid boundary of eight components.

10. The specific geometric message of this format of idol of Lord Brahma, Creator the supreme, is that Lord Brahma is the presiding deity of 4-space with hyper cube 4 as its representative regular body manifesting as of four folds with 2-space in the role of dimensional order, 3-space in the role of boundary, 4-space in the role of domain and 5-space in the role of origin fold.

11. Like that, it would be a blissful exercise for the *sadkhas* (including senior *sadkhas*) to compare their transcendence experiences and as well as their intellectual attainment in the light of the experiences of senior most *sadkhas* preserved in the scripture about higher spaces reality particularly of four, five and six and seven spaces with Lord Brahma, Lord Shiv and Lord Vishnu being the presiding deity of 4 / [symbol], 5 / [symbol] and 6-space / [symbol] respectively and Sapt rishi / [symbol] being the presiding deities of seven space of unity state order lively as the pole star domain.

12. The whole range of Vedic literature, beginning with the very first letter of alphabet is of specific form, frame, format and frequencies of geometric formats, and as such the *sadkhas* shall perfect their intelligence and unify their consciousness to have the bliss of the Vedic systems and orders of which features is the Vedic mathematics. ■

6
TWELVE BASIC FEATURES

1. Short duration twelve weeks course of twelve basic features for selfevaluation of Reality comprehension by the teachers has been designed and same is being made available for facility of all those requiring it for their selfevaluation.
2. The twelve basic features of Reality in terms of whose comprehension one may have selfevaluation are as follows:

   The **first feature** is about the Reality of four space. The **second feature** is about the distinguishing features of this space vis-à-vis our three space, whose chase is aimed to be had by comparative study of representative regular bodies of three space and four space respectively.

   The **third feature** is about the manifestation formats of the entities around us. **Fourth feature** is going to be concerning the origin folds compactification at origin of four space as seat of five space.

   Fifth feature shall be focusing upon the Reality of real 5-space as **Consciousness field** emerging within the intelligence field, while **sixth feature** would be concerning fulfilling of four space domain with solid quantifiers fountained from within the origin of four space.

Seventh feature of this chase is going to be about the manifestations of solid quantifiers and the transcendental range sealed at the origin fold. **Eighth feature** shall be the chase of emergence of transcendental worlds (5-space of solid order) within Creator's space (4-space of spatial order).

Ninth feature shall be a chase of synthesis of manifestation layers with transcendental base. This feature also shall be chasing the transcendental phenomenon of the origin fold acquiring new rule of dimension fold of next manifestation layer because of the transcendental base beneath the origin fold. With it, would emerge the phase and stage of **tenth feature** of transition and transformation from transcendental worlds to transcendental skies manifesting at dimension of dimension level.

Eleventh feature shall be about glimpsing of the transcendental phenomenon of the skylines partitioning the universe into a pair of hemispheres, on either side of the skylines as distinct pair of universes. Finally there would emerge as **twelfth feature**, the phase and stage of transition and transformation for the transcending mind glimpsing at core of the transcendental worlds (5-space) the phenomenon of selfreferral domains (6-space) being lively as the orb of the Sun / Being / Self / Divya Pursha.

3. Each feature is being covered during a week as seven steps chase, with one step a day and thereby the coverage for all the feature is being had within three months as 84 steps.
4. This Vedic geometry course of its all 84 steps covering these 12 features is the subject matter of book-5 of the present series. Here below are enlisted these 84 aspects:

First Feature

Four Space Reality

DAY 1	Introduction to 4-space Reality
DAY 2	First approach to 4-space Reality
DAY 3	Second approach to 4-space Reality
DAY 4	Third approach to 4-space Reality
DAY 5	Fourth approach to 4-space Reality
DAY 6	Fifth approach to 4-space Reality
DAY 7	Mathematics, science and technology of 4-space

Second Feature

Distinguishing Features of Hyper Cube 4

DAY 8	First distinguishing feature
DAY 9	Second distinguishing feature
DAY 10	Third distinguishing feature
DAY 11	Fourth distinguishing feature
DAY 12	Fifth distinguishing feature
DAY 13	Sixth distinguishing feature
DAY 14	Seventh distinguishing feature

Third Feature

Manifestation Format

DAY 15	Four spatial dimensions
DAY 16	Four steps long measuring rod
DAY 17	Four fold manifestation format
DAY 18	First fold also accepts full manifestation format
DAY 19	Second fold also accepts full manifestation format
DAY 20	Third fold also accepts full manifestation format
DAY 21	Fourth fold also accepts full manifestation format

Fourth Feature

Compactification of Origins

DAY 22	Compactification of origins at transcendental core of Creator's space
DAY 23	Chase of as spatial order coiling around transcendental
DAY 24	Solid order of transcendental core supplying spiral formats for spatial coils
DAY 25	Spatial order accepting splits for the spirals around the transcendental stem
DAY 26	The split deserves to be chased along both axes of spatial order
DAY 27	The pair of spatial spirals may be parallel or cross
DAY 28	The spirals may be parallel or cross

and that to, to be in identical or opposite orientations

Fifth Feature

Reality of 5-space Within 4-space

DAY 29	Creator's space supplies manifestation formats
DAY 30	Each quarter of manifestation format may be of expression of full manifestation format
DAY 31	Creator's space makes all spaces and their bodies just being the manifestation layers
DAY 32	Creator's space itself accepts manifestation layer feature for its entities and phenomenon
DAY 33	Solid order of transcendental worlds fulfils Creator's space as solid manifestation layer
DAY 34	The spatial and solid manifestation layers gets superimposed upon each other
DAY 35	The phenomenon of emergence of transcendental worlds within Creator's space deserves to be chased as a phenomenon of superimposed layers, and as a consequence there of, both symmetric and asymmetric set ups co-exist.

Sixth Feature

Fulfilling Solid Quantifiers Within Creator's Space

DAY 36	5-space deserves to be chased as space supplying five steps long measuring rod
DAY 37	5-space deserves to be chased as space supplying five steps long transcendence range for four fold manifestation layers
DAY 38	Each of five steps of transcendence range to accept full transcendence range for transcendence there form
DAY 39	The dimensional order of each step of transcendental range as wellbeing of full transcendence range
DAY 40	The boundary fold of each step of transcendental range as wellbeing of full transcendence range
DAY 41	The origin fold of each step of transcendental range as wellbeing of full transcendence range
DAY 42	The phenomenon of firm establishment of solid quantifiers within Creator's space deserves to be chased fully

Seventh Feature

Manifestations of Solid Quantifiers

DAY 43	Chase real 5-space within Creator's space as Consciousness field within intelligence field
DAY 44	Chase Consciousness field within intelligence field as solid order superimposed upon spatial order

DAY 45	Chase this phenomenon as consciousness impulses superimposed upon intellectual waves
DAY 46	Chase intellectual waves as carriers of manifested values
DAY 47	Chase consciousness impulses as carriers for life virtues
DAY 48	Chase life virtue riding transcendental carriers of rays of the Sun
DAY 49	Chase unity state of consciousness

Eighth Feature

Emergence of Transcendental Worlds Within Creator's Space

DAY 50	Unity state of consciousness leads to transcendental order
DAY 51	Transcendental order leads to transcendence within domain as well as within dimension, to their skylines
DAY 52	Chase skylines of unity state along its first transcendental dimension
DAY 53	Chase skylines of unity state along its second transcendental dimension
DAY 54	Chase skylines of unity state along its third transcendental dimension
DAY 55	Chase skylines of unity state along its fourth transcendental dimension
DAY 56	Chase skylines of unity state along its fifth transcendental dimension

Ninth Feature

Synthesize of Manifestation Layer

DAY 57	Skyline's emergence
DAY 58	Skylines along spatial axes
DAY 59	Ascendance through inner folds of the transcendental worlds
DAY 60	Ascendance along both the axes of the spatial order
DAY 61	Ascendance is to be looked for reversal of transcendence
DAY 62	Cosmic phenomenon of trans sustained by Transcendental Skylines
DAY 63	Coordination of spatial words by the transcendental core of Creator's space

Tenth Feature

Partitioning of Universe Around

Transcendental Skies

DAY 64	Transcendental phenomenon of emergence of transcendental worlds in pairs within Creator's space
DAY 65	Full transcendental world within First hemisphere
DAY 66	Full transcendental world within Second hemisphere
DAY 67	Sustenance of hemispheres by spatial order
DAY 68	Hemisphere along first axis of spatial order

DAY 69	Hemisphere along second axis of spatial order
DAY 70	Partition of Universe as hemispheres

Eleventh Feature

Chasing Distinct Universes

DAY 71	Artifice six and six steps long ranges
DAY 72	Shad Chakras format of Human frame: As external characteristics of measuring rod
DAY 73	Ten Chakras format: Transition from 3-space to 4-space
DAY 74	Five basic elements space and Sun: As measuring rod
DAY 75	Five basic elements space and Sun: As Dimensional order
DAY 76	Five senses band and mind: As measuring rod
DAY 77	Five senses band and mind: As Dimensional measures

Twelfth Feature

Chasing SelfReferrals Domains

DAY 78	Five basic elements range
DAY 79	Divya Pursha
DAY 80	Atman (Self)
DAY 81	Vishnu Lok
DAY 82	Sun
DAY 83	Hyper Cube 6
DAY 84	6-space

■

SECTION - II

THE FEATURE OF LEARNING

1
INTRODUCTORY

INTERLINKING NOTE

Teaching and learning are inherently intimately connected as both being same subject content centred. The 'art of teaching' is 'teachers centred' and the 'science of learning' is 'students centred' but the technology of 'comprehension and skills' of knowledge for acquisition of 'wisdom' is common for both teachers and students. That is why 'the teachers and students' aiming enlightenment jointly pray simultaneous enlightenment for both, students and teachers jointly engaged and devoted for 'knowledge, wisdom and enlightenment'.

This, this way makes 'teaching' as to be inseparable from 'learning'. Learning while teaching, is one aspect, which makes the enfoldment process of knowledge to be very smooth to be naturally conducive alike for the young minds as well as for the mature brains.

However, the learning essentially is 'students centred', which means 'young minds centred' and the same, naturally is to unfold sequentially but smoothly for natural transition from its one step to the next step.

This, as such, makes the learning process to be availing 'cube' as representative regular body of 3-space as 'first learning facility' and 'human body' as the most intimate to young minds, as 'the second learning facility'.

This pair of learning facilities shall be helping to approach the 'existing mental format of young minds' for transition and transformation for 'next phase and stage mental format of mature brains'. It is attainment of this transition and transformation which is to be taken as a success of learning of this phase and stage of schooling.

The success range of learning of this phase and stage of schooling is to be taken to be as is the range between the starting position of the mental format and the attained state for the mental format at the end of the learning.

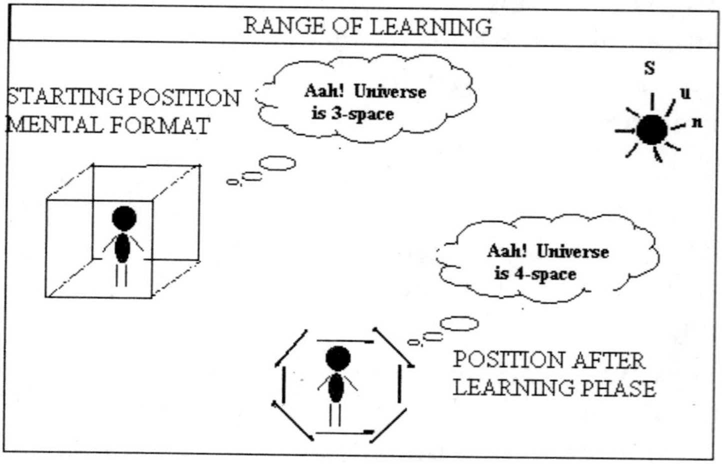

These end points of the range, may better be taken as the *sadkha* itself as the starting point and the guru as the end point. This, in the language of the Discipline of learning, may be described as the start with mental format of the student being as his existence domain being that of a cube and its attainment state being the domain of a hyper cube 4.

INTRODUCTORY

1. Vedic mathematics is the ancient Discipline. The range of the ancient wisdom as subject matter of instructions of Discipline of mathematics is about 'the Creations and the Creator'.
2. 'The creation' domain is designated as 'Triloki / Jagat / Vishwa / 3-space '.
3. The '*Vyasti-Samasti*' order of creation makes out, complete manifestation of total features, characteristics and properties of 3-space in its representative regular body, namely, a 'CUBE'.
4. 'Cube' is the specific / '*Samasti*' form / printout of 'general / *Vyasti* form / domain of 3-space.
5. The Creator's space is designated as Brahma Lok with Lord Brahma / the four headed Lord / Creator the supreme as its presiding deity.
6. Creator's space provides four fold format for manifestation of creations and the same is also availed by creator Himself as a format for its own idol.
7. The format of idol of Lord Brahma is of four heads with each head equipped with a pair of eyes; Lord sitting majestically on a lotus seat of eight petals

and meditating blissfully upon its Lord, Lord Shiv, in cavity of its own heart.

8. This format is the format parallel to the set up of hyper cube 4 as a spatial order 4-space enveloped within a solid boundary of eight components.

Hyper cube-4

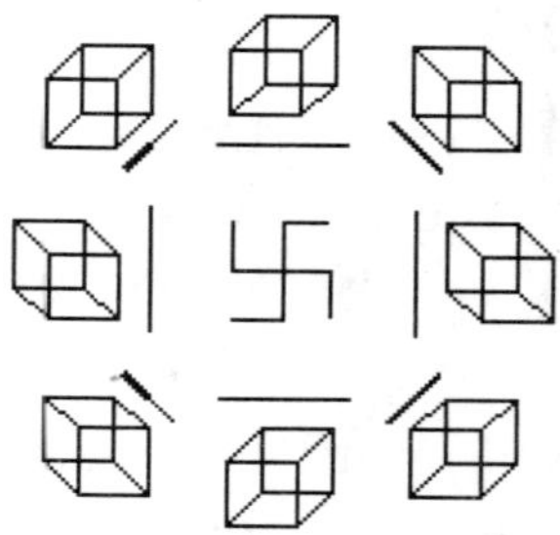

9. As such the learning and teaching of ancient system of Vedic mathematics as art and science of the teaching of Vedic mathematics deserves to be chased beginning with the cube as a Samesti form / printout of 3-space and reaching up till the Vahesti form / format of 3-space domain.

■

Technical terms up to date

'***Vyasti-Samasti***' is complete text of Ganita Sutra 11. The simple English rendering for its working rule and functional format may be taken as 'Specific and general', of which illustrative situation is of 'cube as specific expression and 3-space as general expression'.

■

2

CUBE TO HYPER CUBE-4

1. The learning and teaching range of Vedic mathematics is from cube to hyper cube 4.

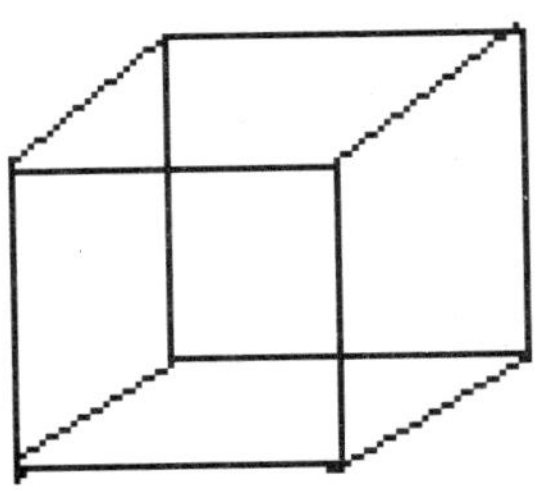

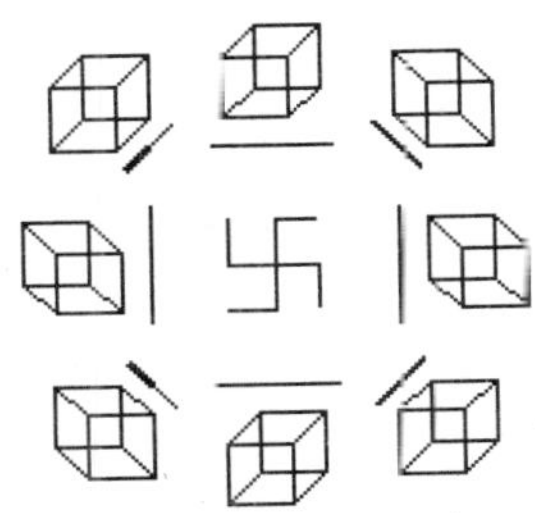

FIRST EXERCISE

2. First exercise is to have a cut of the cube into eight sub cubes to visualize and to have an insight about the synthetic set up of a cube of eight cubes.
3. It would be a physical exercise of slicing a soap cake or of a bread. The slicing is to be had with a knife in three cuts parallel to the axes planes.
4. As a first step, first cut of the soap cake is to be parallel to the vertical plane through the centre of the cube.

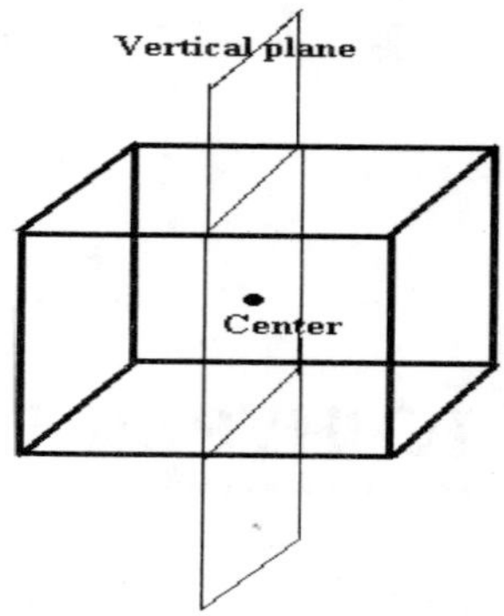

5. The second step, as second cut, is to be along the second vertical plane as follows:

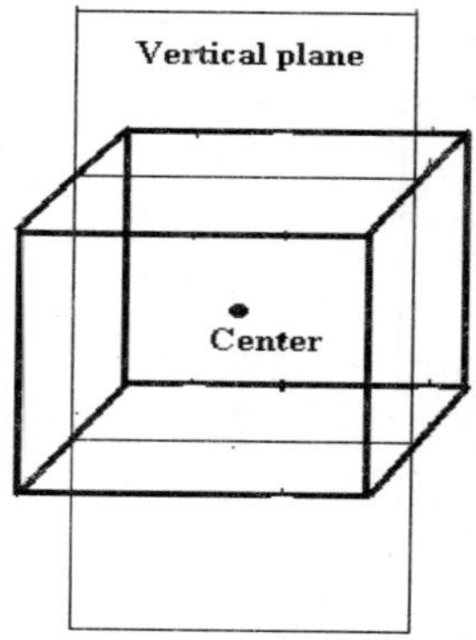

6. The third step, as third cut, is to be along the horizontal planes as follows:

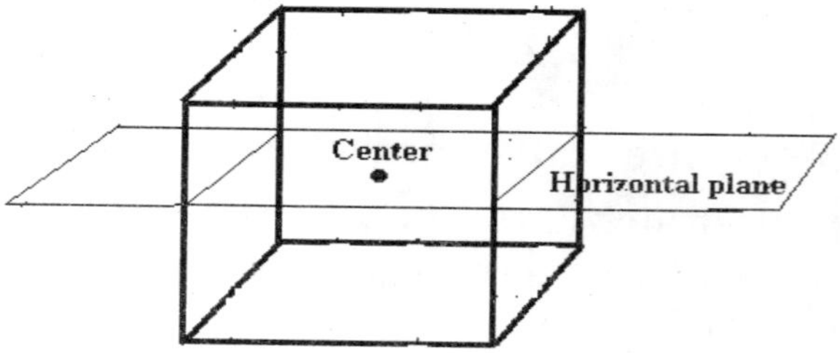

7. The combined effect of all the three steps would be to have a cut for a soup cake / bread / cube into eight parts / sub cubes.

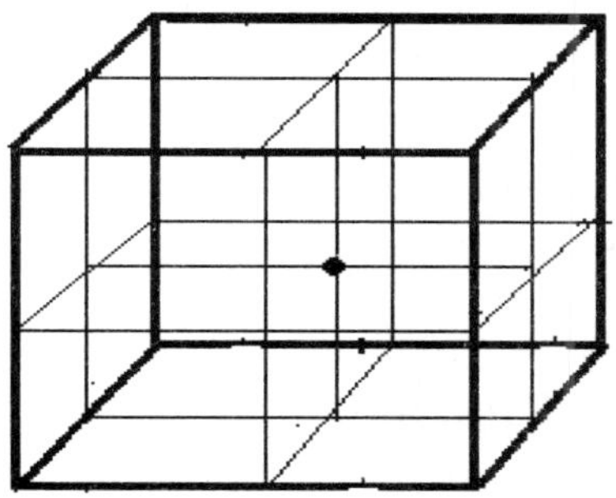

8. 'Centre of the cube' is enveloped within eight sub cubes of the cube.

SECOND EXERCISE

9. The second exercise is to chase a line / interval (1-space body) as a track of a moving point (0-space body).
10. Further to chase surface / square (2-space body) as a track of a moving line (1-space body).
11. Still further to chase solid / cube / 3-space body as a track of moving surface (2-space body).
12. In a sequence a step further shall be taking us to hyper cube4 (4-space body) as a track of a moving solid (3-space body).

THIRD EXERCISE

13. The third exercise is to comprehend as that interval accepts 2 boundary components, namely two points (0-space bodies).
14. Further to comprehend, as that square accepts

four boundary components, namely four lines (1-space bodies).

15. Still further to comprehend as that cube accepts 6 boundary components, namely six surfaces (2-space bodies).
16. In a sequence, it to be comprehended as that hyper cube-4 accepts eight boundary components, namely eight solids (3-space bodies).

CONCLUSION

17. In the light of the above three exercises, one may have insight about the existence of hyper cube4 at zero (point) value at centre of the cube, as a track of a moving solid, and as a domain / space enveloped within eight solid boundary components.
18. Further, one may glimpse a parallel for the set up of hyper cube4 as well as of the features of idol of Lord Brahma, four Head Lord, Creator the Supreme.
19. It shall be a blissful exercise to permit the transcending mind to glimpse centre of the cube being a seat of the format of hyper cube-4 as format of idol of the Creator (presiding deity of Creator's space). ■

Technical terms up to date

Four head lord (Lord Brahma) is the presiding deity of Creator's space (4-spac). Along the format of idol of Creator manifests hyper cube-4. ■

3

TO BE WITHIN A CUBE AND TO BE OUT OF IT

1. To be within a *SHILA* / slab, a stone slab, a solid slab, a cube, and out of it, has been the specific explorative chase of the ancient wisdom and it is

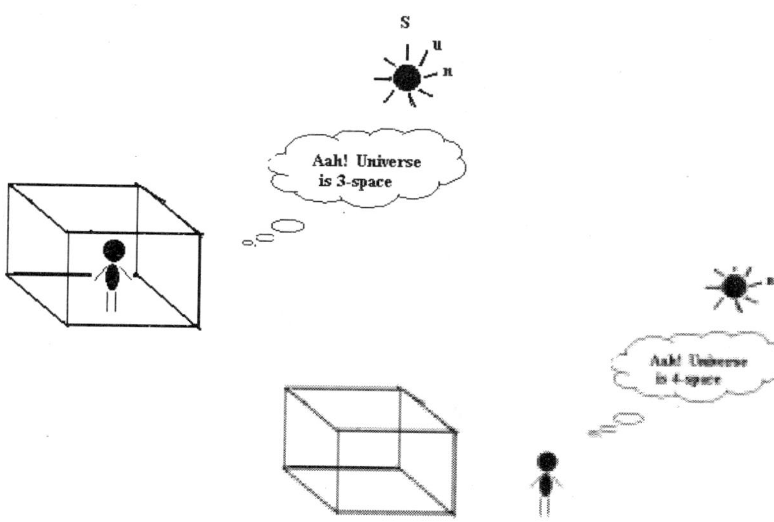

a chase to be followed by the *sadkhas* and students of the ancient discipline of Vedic mathematics.

2. The chase on the lines is to put oneself in a

position of existence within a cube, and at the phase and stage of out of it as to be face to face with the set up of a hyper cube4.

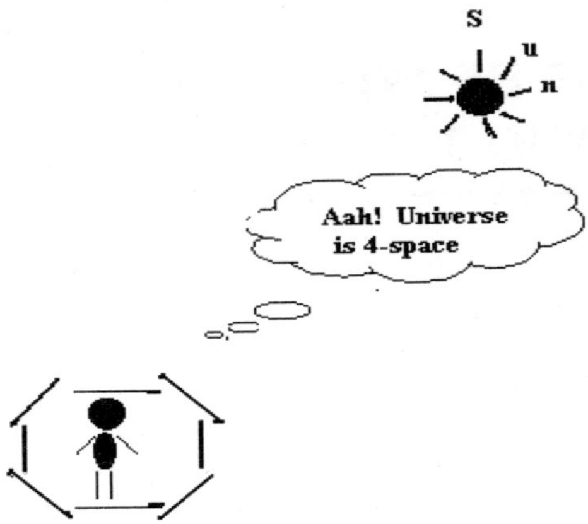

3. The phase and stage ahead would be when one shall be taking oneself out of the 4-space cage and would be within the mental frame of comprehension of the universe as 5-space.

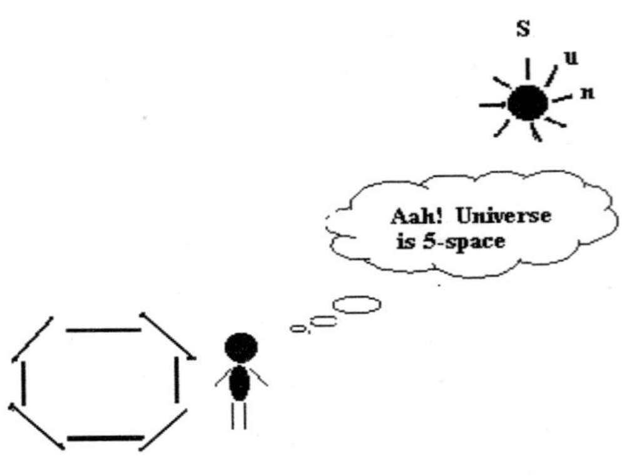

4. This is a phase and stage of the transcendental worlds described in the scripture as the phase and stage of the Creator, Lord Brahma, the four

head Lord meditating in the cavity of his heart upon his Lord, Lord Shiv, the five head Lord, the presiding deity of the transcendental worlds (5-space)accepting manifestation format for its idol within Creator's space as of hyper cube-5 set up with its domain boundary ratio as A^5 : 10 B^4, which as representative regular body of 5-space, that is, as hyper cube5, may be symbolized as is indicated.

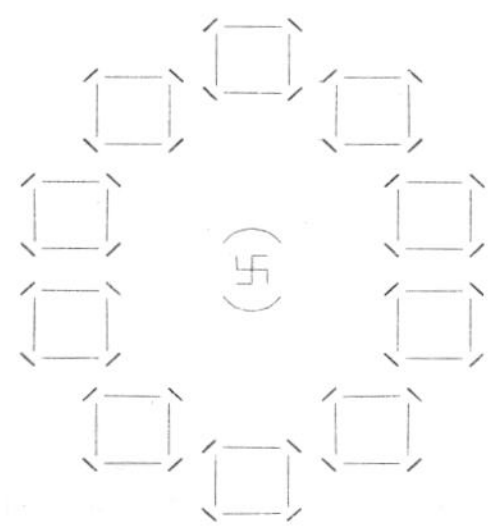

5. Scriptures describe this phase and stage of the Creator (4-space) multiplying ten fold, as ten Brahmas in Shiv lok at manifested boundary of the transcendental worlds.
6. A chase ahead shall be taking us to hyper cube-6 with a domain boundary ration being A^6 : 12 B ^{5}

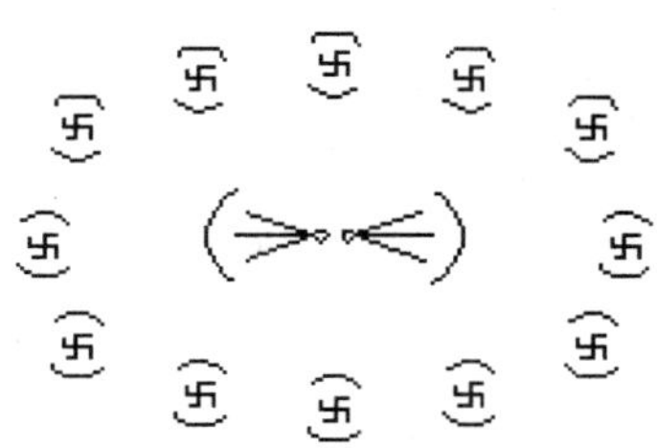

HYPER CUBE-6

■

4

FIRST BOOK OF VEDIC MATHEMATICS

1. The first, and the source book of Vedic mathematics is the 'human body' and it is to be 'read' beginning as existence domain framed within a cube and by being out of it to reach at its second, that is, the micro state of existence phenomenon within human frame availing 4-space domain.
2. The young minds deserve to be exposed to the discipline of this ancient wisdom of Vedic mathematics expression in terms of the human body as the best gift brought by the child straight from the abode of the Creator.
3. The beauty of this gift is that it is to be the creator's book of instruction for the whole range of the existence phenomenon from the first moment of life till the last moment of the life.
4. As such the natural way for learning and teaching Vedic mathematics would be to acquire the art and science of reading the commands of the Creator as are manifested in the human body of its full range beginning with its outer most fold as of its form, frame format.

5. The first lesson for initiations for the young minds is to be to help them to start reading this Creator's book of mathematics by chasing it as its features of set ups reflecting as symmetry of the body.
6. It is 'this' left and right 'pair' of reflection images set up around the line of symmetry marking as the central nose line on its front face and as a back bone line at its back.

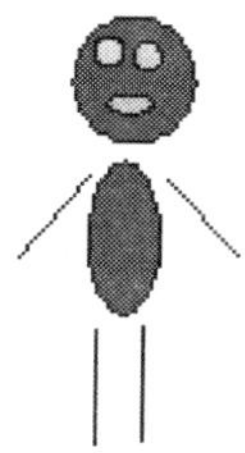

Left -Right set up as reflection pair

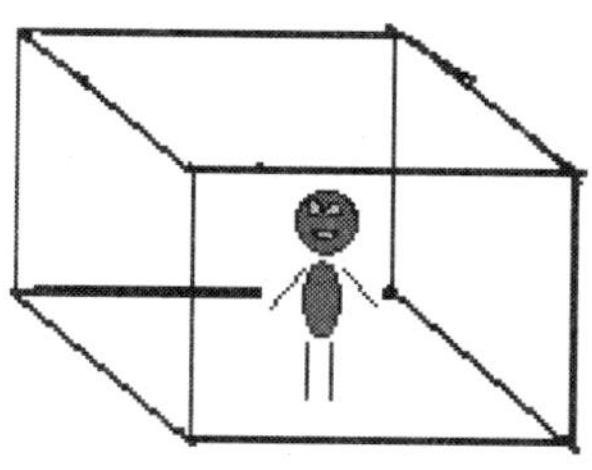

7. Beginning like that, the young minds are to be gently, as a set of sequential steps, to be taken to the following pages of 'this Creator's book' of mathematics by taking it as the start with existence domain to be as framed body is meant for protection of its outer fold while being coordinated in terms of the existence domain 'within a cube'.
8. Sequentially, for the next phase and stage of

existence at micro state (Shuksam Sharir), the young minds shall be smoothly initiated for required transition from the existence domain within a cube to that of the existence domain within a hyper cube4.

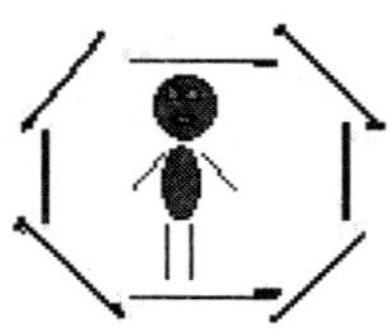

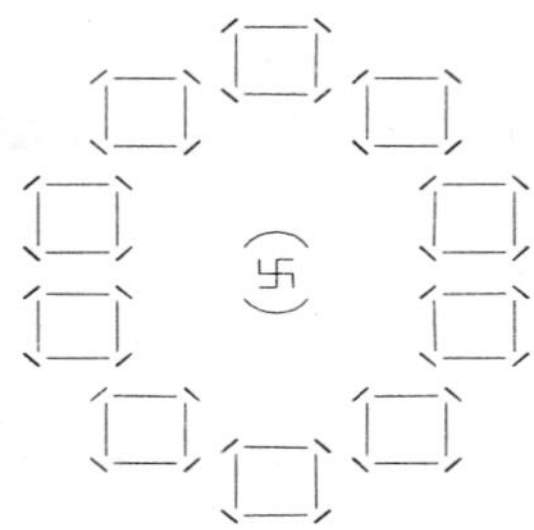

9. Then would follow the phase and stage of further transition and transformation for take off from the functional domain of Creator's space within hyper cube4 to that of the functional domain of transcendental worlds within hyper cube5.

■

5

VYASHTI-SMASHTI EXPRESSIONS

1. Young minds are to be very gently impressed upon the phenomenon of expressions for existence domains.
2. The Samashti expression of the 'domain' may be exposed as 'specific expression' of the domain while the Vyashti expression being the general expression of the domain.
3. This fine concept being of a very delicate feature, the same as such shall be exposed in a sequential steps availing the well impressed upon illustrative situations like that of the balls of finite and infinite sizes, lines / intervals of finite and infinite lengths, surfaces / squares of finite and infinite areas and solids / cubes of finite and infinite volumes.
4. All finite balls, lines / intervals, surfaces / square and solids / cubes are to be impressed upon as the specific expressions of infinite domains of space, along one axis, pair of axes and triple axes set ups.
5. Very gently it is to be impressed upon the young

minds as to the feature of all finite intervals being of same set of properties, and as such all these being the specific (*Samasti*) expressions of 1-space.

6. Likewise, the young minds be smoothly taken through this *Samasti* / specific expression feature of 2-space in terms of all finite areas / surfaces / squares being of identical set ups. This exercise, at a next step is to be carried forward for all finite volume cubes being the *Samasti* / specific expressions of 3-space and the same being of identical set of features.
7. For facility of the young minds the symbolic / figurative / symbolic representations of a pair of intervals of different lengths (and, likewise a pair of square of different areas and also a pair of cubes of different volumes) may be taken up, step by for their 'one to one' correspondence of 'point constituents' of intervals (and of line constituents of squares, and square constituents of the cubes).

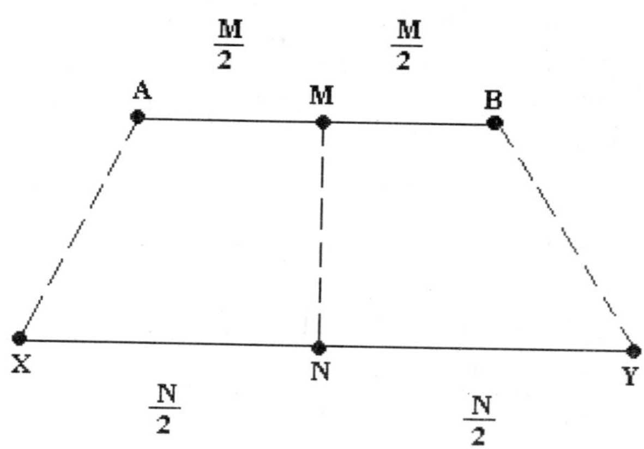

8. This, for facility of young minds, be chased step by step, may be on the following lines.

   Step 1. Let the pair of finite intervals of lengths M and N be drawn as depicted above.

   Step 2. The end A of the first interval be joined with end X of second interval. Likewise end B of the first interval be joined with end Y of the second interval.

   Step 3. The middle M of the first interval of length M be joined with middle N of the second interval of length N.

   Step 4. Likewise the middle of first half of first interval, namely A M to be joined with middle of the first half of the second interval namely XN. The same way, the middles of second halves of both intervals as well to be joined.

   Step 5. This sequence of step 4 above, now to be repeated for two halves of the first halves of both intervals.

   Step 6. This process would continue infinitely as 'the length' in its process to be 'halved' is never to extinguish as to be of value of 0 length. And, as such, both intervals A B and X Y, are to be of identical features of specific / Samashti expression of 1-space.

   Step 7. Now, when both A B and X Y would go infinite with their both ends for both of them to be at infinity, the same would become identical Vaheshti / general expressions for 1-space.

9. Then, at the next step of reaching at comprehensions of Sameshti and Vyashti

expressions of 2-space for the squares of finite areas, and for squares of infinite area, the parallel steps on above lines may be availed.

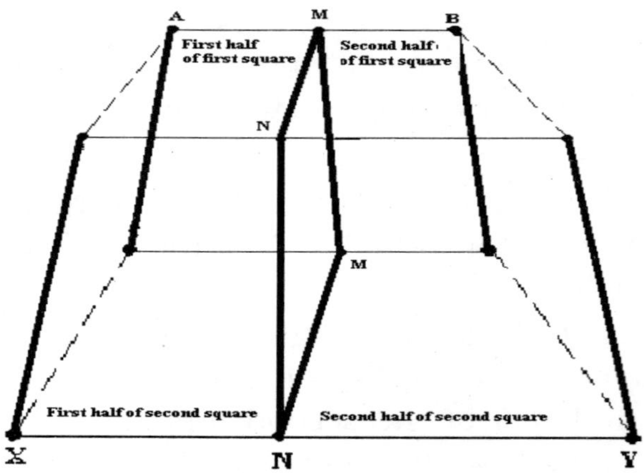

10. This way, at next step for Sameshti and Vyashti expression for cubes and 3-space, the domain of 4-space may be availed.
11. And, for Sameshti and Vyashti expressions for hyper cube4 and 4-space, the domain of transcendental worlds (5-space) need be availed.

■

6
SETTLEMENT OF BASICS

1. The settlement of basics of Vedic mathematics is to be had in a Creator's space, that is, in 4-space.
2. This, this way, would mean that the processing for the purpose is to be had beginning with the set up of a cube within 4-space domain.
3. First, as such, it would take us to the different set ups in which there is simultaneous availability of 3-space and 4-space set ups.
4. One such set up is of hyper cube 4, as a four space domain enveloped within a solid boundary, with domain boundary ratio formulation being $A^4: 8\ B^3$.

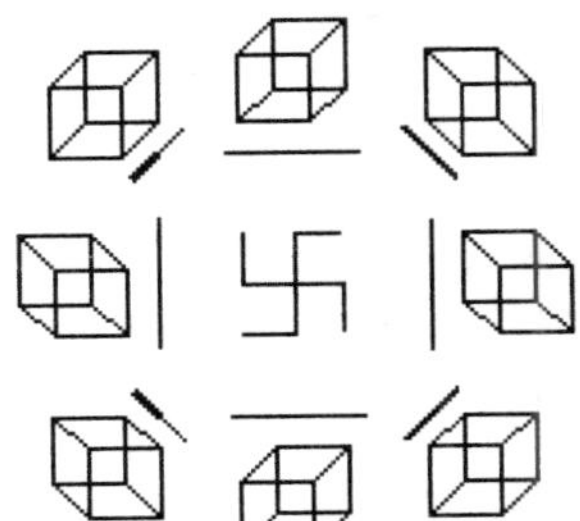

5. The manifestation format supplied by Creator's space accepted by idol of Creator itself, is of four folds.

   Firstly as of four heads with pair of eyes in each head / 4 spatial dimensions (2-space in the role of boundary),

   Secondly the lotus seat of eight petals / eight solid boundary components (3-space in the role of dimension)

   Thirdly the 4-space as domain

   and **Fourthly** 5-space in the role of origin / cavity of heart of Lord Brahma being the seat of his Lord, Lord Shiv, the five head Lord with three eyes in each of the five heads (5-space, the solid dimensional space in the role of origin fold of 4-space).

6. This manifestation of transcendental worlds (5-space) within Creator's space, manifests its origin as a seat of solid order 5-space. With it, 3-space plays role of dimension of the origin fold of 4-space.

7. Further, 3-space within 4-space domain, is having one degree of freedom of motion along fourth dimension, and as such the 3-space acquires dynamic state within a 4-space.

8. In addition, the manifestation of cube, within 4-space, is also of four folds, with 1-space being in the role of dimension, 2-space in the role of boundary, 3-space as domain and 4-space in the role of origin of 3-space.

9. This leads to a geometric envelope for 3-space content lump / volume / Domain / domain fold, a synthetic set up of eight corner points, twelve

edges and six surface plate, together being a set up of twenty six geometric entities.

10. The young minds, as such are to be gently exposed to this manifestation set up, as well as of synthetic set up of cube and its envelope.

11. This manifestation set up and the synthetic set up of cube and its envelope are to be chased to reach at the basics of creations within Creator's space as basics basis of Vedic mathematics.

12. From static position to that of dynamic state of the cube within four and higher spaces is one pair of set ups to be chased.

13. The fixation of points within the domain of cube in reference to its envelope, and the fixation of the envelope itself in terms of the envelope of the hyper cube4, is another pair of set ups to be chased.

14. The Creator's space as origin source reservoir for the cube, and the Creator's space itself being of the privilege of transcendental origin source reservoir, is another pair of set ups to be chased.

15. The emergence and fading of solid order transcendental fields within Creator's space is other phenomenon to be chased.

16. For this chase of basics of the ancient discipline of Vedic mathematics, young minds need be initiated into the ancient processing technique of transcending mind to transcend through and chase the transcendental phenomenon of sequential emergence of transcendental fields.

17. As such the learning and teaching of Vedic mathematics beginning with settlement of its

basics' basis, the ancient technique of transcendence of the mind and glimpsing of the inner folds of the transcendental worlds is to be availed and the *sadkhas* and students fulfilled with intensity of urge to fully know and to completely chase the basics' basis of ancient discipline on first principles shall be convinced for the blissful fruits of prolonged sittings of trans.

A fresh look at the set-up of the cube in the light of following figures with focus respectively upon corner points, edges and surfaces shall be helping to have a comprehension view of the geometric envelope of a 'cube' .

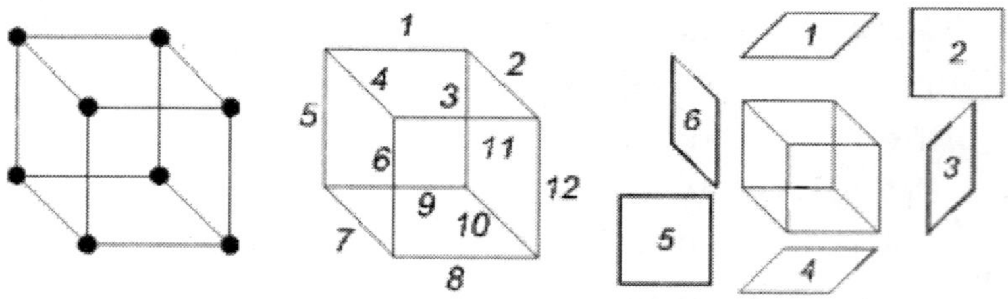

■

7

INITIAL SKILLS FOR YOUNG MINDS

1. At every opportunity, a young mind be chiseled for fresh look at the set up of the cube for initiations for perfection of intelligence of the order of the perfection of the formats of creations.
2. Young minds shall be exposed to the feature of the set up of the cube of proper fixation of all the eight corner points' equidistance from the center of the cube.

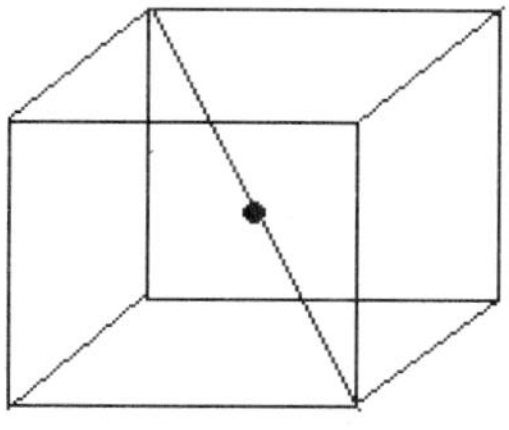

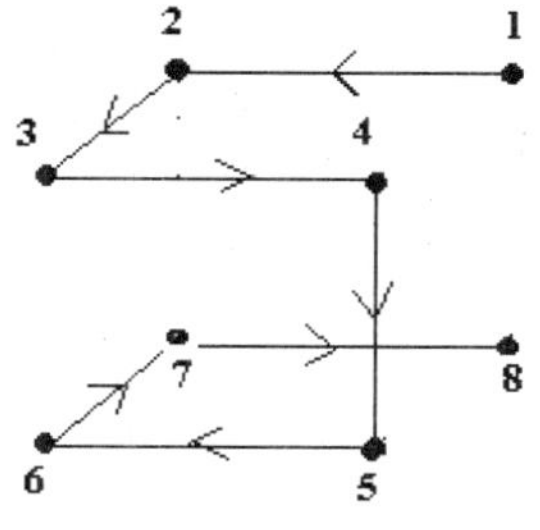

3. Young minds also be taken through the set up of the cube coordinating all the eight corner points in terms of seven edges arranged in continuing orientation flow beginning from any given corner and leading up till the eighth corner.

4. Young minds shall have a fresh look at the set up of the cube as that its every edge can be approached for whole of it from either of its end points as corner of the cube.
5. The continuous orientation flow from any corner of the cube up till its eighth corner is to accept reversal of orientation accepting eighth corner of original sequence as the starting point and the first corner of original sequence to be the last point of reverse orientation.
6. The eight corner points and twelve edges are of values in the ratio 1:3/2.
7. The eight corner points and six surfaces of the cube are of values in ratio 1: ¾.
8. The twelve edges and six surfaces of the cube are of values in ratio 2:1.
9. The eight corner points, twelve edges and six surfaces of the cube are of the values in ratio 4: 6:3 with middle value (6) being double of the value (3) on its one side and being of One and a half times of the value (4) on its other side.
10. Each of the corner points is embedded with a 3 dimensional frame of half-axes.
11. The 3 dimensional frame of full axes is embedded at centre of the cube.
12. The 3 dimensional frame at centre of the cube accepts a split into a pair of 3 dimensional frames of half-axes along orientations of each of its three axes.
13. The line passing through the centre and joining a pair of corner points with one such corner points

being on either side of the centre, when being one of the axes of the 3 dimensional frame at the centre of the cube, then the split up of this 3 dimension into a pair of three dimension frame into half dimension, these shall be of opposite orientations not only in respect of each other but also in respect of the 3 dimensional frame of half dimension embedded into the given pair of corner points of the cube along the line which connects them as passes through the centre.

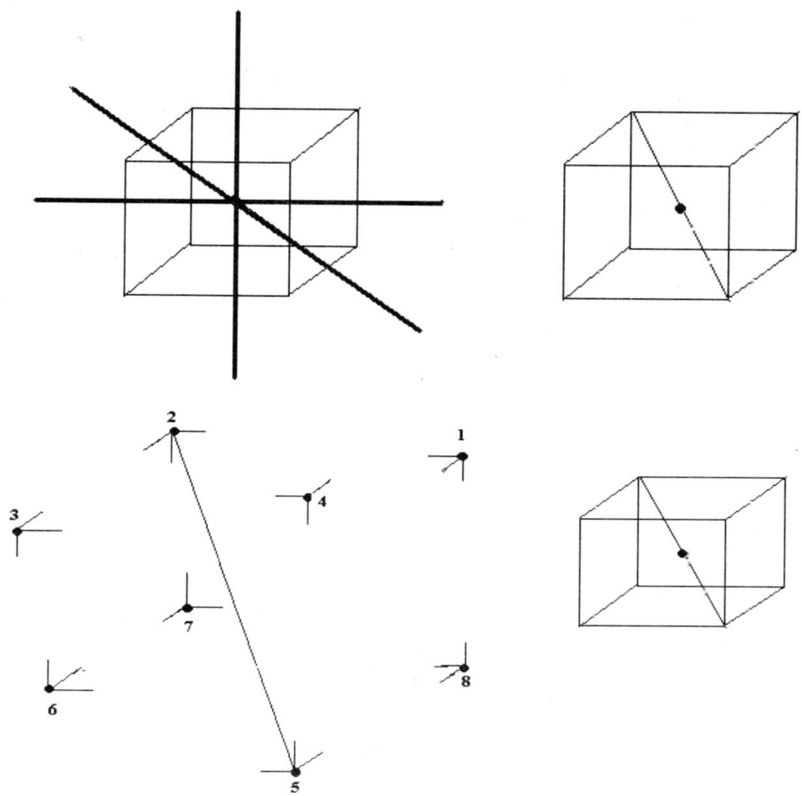

14. Young minds be encouraged to have a fresh look at the set up of the cube, as many times, as one can and at every such opportunity, a fresh expression of comprehensions about the set up of cube be attempted for its evaluation by the teachers having accepted responsibility for helping the students for perfection of their intelligence.
15. One may be sure about ones perfection of intelligence only with the initial stage perfection of comprehension of the features of the set up of the cube at it is only after attaining this phase and stage of perfection, that one may be encouraged to have a take off for the next phase and stage of attainment of unity state of consciousness which as a field emergence within the field of intelligence itself.

A word with the teachers

Teachers may avail here an opportunity to impress upon the young minds as that the set-up of human frame is such that firstly it creates a SENSORY FIELD in terms of an inflow through the interaction of senses. Then, secondly, it creates an intelligence field with transcendence of mind through the solid boundary of Creator's space. And finally, with glimpsing of the transcendental core of the Creator's space their emerges of consciousness field within the intelligence field itself.

■

8

CHASE ON FIRST PRINCIPLES

1. Young minds be very gently exposed for chase on first principles of basis of basics of Vedic mathematics in terms of working rules of Ganita Sutras and the set ups of cube and of the human frame as first book of mathematics about creations within Creator's space (4-space).
2. The Creator's space is of spatial order, while the cube manifests its set up within Creator's space along its manifestation format as of a linear dimensional order.
3. It is this delicate feature of the manifestation aspect of our Triloki / Vishwa / Jagat / 3-space / cube being of linear order manifesting along the spatial order, that every care deserves to be taken while young minds are exposed to these delicate characteristics of emergence of linear order within Creator's space.
4. In fact this emergence of linear order within Creator's space is the transcendental skyline / dimension of dimension of the transcendental worlds.

5. As such, the phenomenon of emergence of linear order and hence manifestation of the set up of solids / cubes is essentially the manifestation phenomenon at transcendental skylines of the transcendental worlds within Creator's space.
6. However, when the set ups of the cubes are to be taken up in isolation for their comprehension and chase, then the fact that there would be intentional cut off from the base domain of 4-space as source origin of 3-space set up and further also there would be a swearnace from the transcendental worlds basis of the base of Creator's space, and as such, the comprehension chase of the set up of the cube is bound to be of such limitation and restriction about their full and complete chase in the process.
7. But as the set up of cube is of perfection feature of the intelligence field of the Creator's space domain, which is of spatial order, as well as of the consciousness field emerging within intelligence field as being of solid dimensional order, as such the repeated states of such fields shall be ensuring the unity state features for the set up of the cube as well which as such shall be coming within the comprehension of the glimpsing range of the transcending mind despite the cut off and take off of the cube from its base and basis at the time of its chase comprehension.
8. Nevertheless, despite the par-excellence of the set up of the cube, the young minds certainly would be requiring very soft carrying along the pointed chase paths and accordingly deepens the

responsibility of the teachers to sequentially face the transition step for smooth transformation and perfection of comprehension for the so initiated young minds.

9. The Ganita Sutras, in the sequence and order of their setting, provide us the set of functional rules which are to be availed for chase on first principles of basis of basics of the ancient wisdom of Vedic mathematics.
10. As such the young minds are to be very gently initiated into the functional rules and corresponding skills of the Ganita Sutras in the sequence and order of their setting from first to the last, that is 16 Sutras and simultaneously from first to the last that is 13 Upsutras.
11. Of these the functional rule and skill of the first Ganita Sutras, naturally is to be of initial focus and attention, as it as the beginning of the whole system, is to make the system to control itself as of features which to carry us up till the other end of the system itself.
12. This selfreferral and selfsustaining potentialities of the processing system of the Ganita Sutras is inherently inbuilt in the system itself because of its availing of the formats within Creator's space. The way this feature prevails throughout the text-scripture of Ganita Sutras deserves to be chased step by step in the order and sequence of setting of the Sutras 1 to 16 and Upsutras 1 to 13.

■

SECTION - III
SUBJECT CONTENT

1
INTRODUCTORY

The 'teaching and learning' of Vedic mathematics, and attainments of it, together constitute one aspect of this ancient Discipline of knowledge. These features are, in fact the 'beginning and end' phase of the course schedule. The central core of this chase is the actual 'subject content'. It as well is of dual features, parallel to the 'pure and applied values' of the ancient wisdom.

This central core is 'subject content centred'. It is designated as 'basics and skills'. Book-3 and 4 are aimed to cover it. Book-3 covers the 'basics of Vedic mathematics'. Book-4 is to cover the skills of Vedic mathematics for attaining perfection of intelligence.

The present phase and stage of book-3 'Vedic mathematics basics' takes to 'basis of basics' for their coverage on first principles. This, as such shall be taking us to the 'set up of a cube' as the package of 'basics' with centre of the cube being the 'single point' source reservoir as origin of 3-space, and the seat of Creator's space (4-space).

As such, the whole focus here for chase of basics at their basics on first principles is to be to follow the set up of the cube for comprehension of its basics and basis and in the process to attain perfection of intelligence.

The beauty of the par-excellence systems of Vedic mathematics lies in the features in built in the system meant for the process of perfection of intelligence to run parallel to the comprehension of the set up of the cube, as a solid boundary of the Creator's space which on its attainment becomes lively as intelligence field.

The further chase of the set up of hyper cube4 / 4-space domain would amount to attainment of consciousness field emerging within the intelligence field. The students with attainments of this phase and stage of learning as perfection of intelligence, would be continuing further learning of next phase and stage for further attainments of the order of a unity of consciousness.

For attainments of chase of basics at their basis on first principles, and for attainment of perfection of intelligence, this phase and stage of learning is offered with best wishes for the *sadkhas* fulfilled with intensity of urge to fully comprehend and to chase the basics basis on first principles.

■

2
BASICS REACHED AT

Being a start with position for initiation of the learning for the middle school levels students, and the learning being aimed to be on first principles availing basis of basics of mathematics as chased by the ancient wisdom discipline of Vedic mathematics, as such, responsibility lies upon the teachers to gently expose the young minds to the delicate features of their this phase and stage of learning. This lesson content, is in fact meant for sharing with teachers, the aspects of the 'path of learning through which the learners have to be taken through' as along this path they have to carry the young minds along with them. As such, the lesson is being numbered as lesson 00, as, the same is being shared with the teachers.

1. First learning is to be about the 'basics already reached at'. The teaching on 'first principles' is to avail the 'basis' of these 'basics'.
2. As such, from learning point of view, the whole focus is to centre around the 'basics' of the 'Discipline' to be learnt.
3. The 'Discipline' here being the 'Vedic

mathematics', as such, the first attention is to be about the 'basics of Vedic mathematics', as a subject content, as well as the process of learning of the subject in terms of 'these basics'.

4. Vedic mathematics being of the unified range of 'ancient wisdom' about the 'existence phenomenon' in its entirety, as a 'single discipline' for the whole range of knowledge, as such, while learning this Discipline through its basics, one is to every remain conscious as that there is a very wide range of applied values of this Discipline, but that, at initial learning stages, focus is to be guided for only the required 'applied values' for the desired skills of 'technologies'.
5. Ancient wisdom approaches the existence phenomenon from its affine state expression of all creations all around fulfilling the universe in the setting in the affine setting thereof as it is.
6. Taking 'affine setting of creations' as these are, as the starting position, the next step of basics is about the forms, frames and formats acceptable by the settings of the creations and the orders which inherently sustain this setting as it is for the natural flow of life and the existence phenomenon.
7. One such order deciphered and chased by the ancient wisdom is focusing around the position of the Sun in this setting, as a focal point of the Earth based existence phenomenon. This chase leads to as that the 'basics reached at' attain the parallel setting of existence phenomenon within the human frame.

8. 'The basics of sustenance of affine state setting of existence phenomenon and flow of life within human frame lead to 'the Sun outside in the universe' as focal point of the 'outside universe', and 'the Sun inside within the human frame' as focal point of the 'inside universe', being of parallel features, characteristics and values.
9. It is this 'basis of the basics' of the 'existence phenomenon on Earth' with 'Sun as the focal point', supplied by the 'ancient wisdom' is to be availed for chase of the 'Discipline of Vedic mathematics' on its first principles.
10. 'Earth to Sun' is a six steps long range of 'Earth, Water, Fire, Air, Space and Sun'.
11. The 'Earth' as Vishwa / Jagat / Triloki /3-space is the blissful existence phenomenon sustained as flow of Jyoti within rays of the Sun, as Vishnu lok / 6-space.

'ज्योतिअभ्यनतरे त्रिलोकी मनोहरः'

Jyotirabhyentre Triloki Manoharm

- Braham Vaverat Puran

12. The 'learning' of the 'basics already reached at and being availed' by Vedic mathematics, as an ancient discipline, as preserved in the available scriptures, as such, is to begin with a focus 'to centre around' the existence phenomenon of its affine setting and natural flow of 'life' within human frame.
13. With it, the starting point for this chase emerges to be the set up of the 'cube' as the representative regular body of 3-space / Vishwa / Jagat / Triloki.

14. Beginning with the affine set up of 'Cube', and reaching up till the affine set up of 'hyper cube6' as the representative regular body of 6-space / Sun / Vishnu lok / Atman / Divya Pursha, is to emerge as the range of learning.
15. The Sathapatya Upved scriptures accept a measuring rod constituted by representative regular bodies of 1 to 6 space, and as such the range of learning is to be parallel to the steps of this measuring rod.
16. The symbolic expression for this measuring rod may be as follows as a sequential set up of representative regular bodies of 1 to 6 space, individually being designated as interval, square, cube, hyper cube4, hyper cube5 and hyper cube6. Collectively these bodies may be designated as hyper cubes 1 to 6.

MEASURING ROD

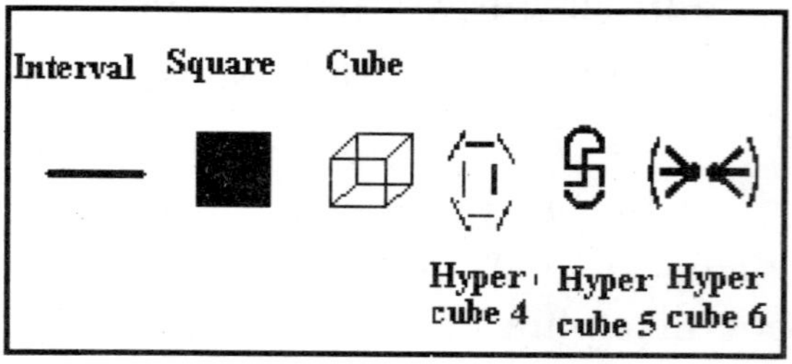

■

3

CUBE AS FIRST FRIEND FROM WORLD OF CREATION

1. 'Cube' () is our 'first friend' from 'the World of Creation'.
2. To know our friend intimately, we may have a fresh look at the affine set up of the cube accepting 8 corners, 12 edges and 6 surfaces.

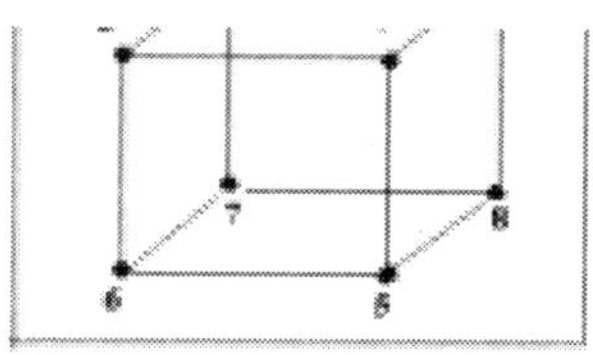

3. Count and number the eight corner points of the cube.

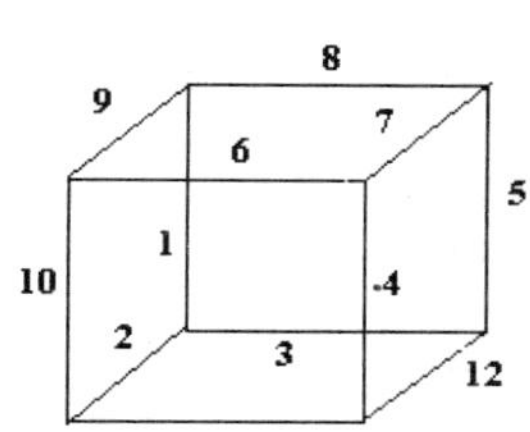

4. Count and number the twelve edges of the cube.

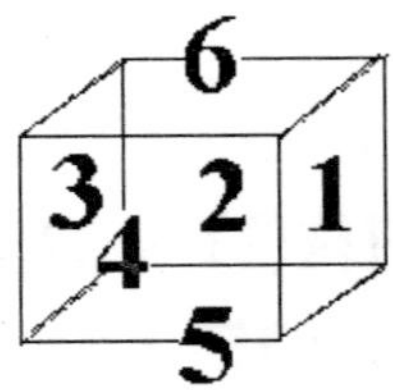

5. Count and number the six surfaces of the Cube.

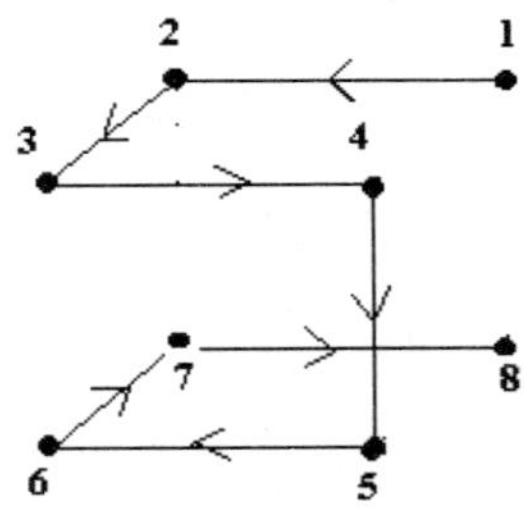

6. One shall get more intimacy with the set up of the Cube, by attempting to connect all the eight corner points in terms of seven edges by maintaining the orientation flow.
7. One shall have a fresh look at the set up of a cube and try to comprehend and expressed more about the set up.
8. One may look at the set up of the cube as a geometric envelope / bag / container. Express yourself about this view of this set up.
9. The view of the cube as an envelope, stitched all around having no window to peep through, puts us in a situation as to how to have more information and knowledge about 'the space content responsible for the volume of the cube' and the connected

features, characteristics and properties of the set up.

10. Express fully as to how intimate you are with your this friend from the 'World of Creation'.

11. 'Cube' to 'Cuboid' setsup sustainable witihin the synthetic boundary stitched in terms of 8 corner points, 12 edges and 6 surfaces deserve to be chased simultaneously as a range from 'CUBE' to 'CUBE END' / 'CUBOID'.

12. It would be refreshing to appreciate as that NVF (CUBOID) = 54= NVF (SUN) in the light of the transcendental manifestation layer accepting 3-space in the role of dimension tagged with origin being played by 6-space.

■

4

THREE FLOW ORIENTATIONS FROM THE CORNER OF A CUBE

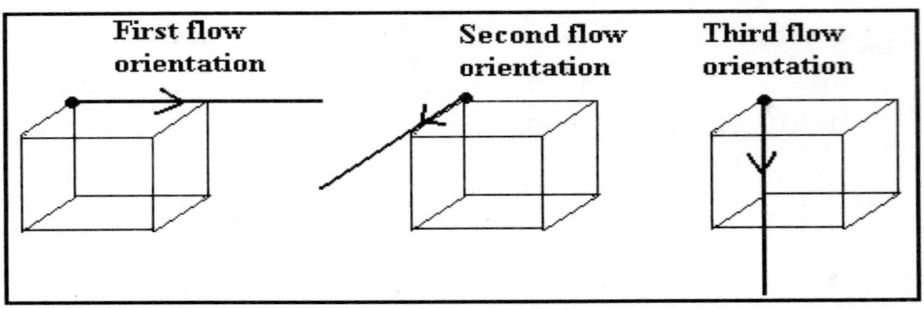

1. One shall see that three edges meet at each corner point of the cube.
2. One may view it as a feature of this set up as that from each corner of a cube their may be three distinct flow orientations along three distinct edges.
3. It would be taken as an exercise to carry forward along the orientation of a given edge to connect all the eight corner points.

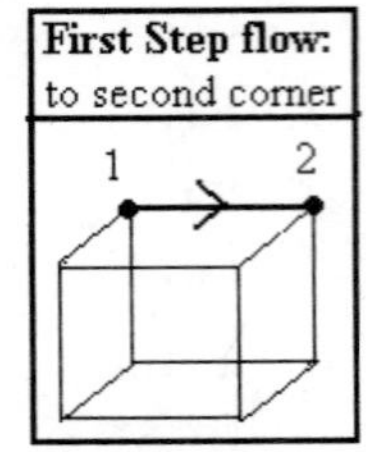

4. Here, one may be confronted with a pair of options at the second step, as is being illustrated here under.

Step 1. To proceed along the first orientation and reach the other end of the edge at second corner of the Cube.

Step 2. At the second corner, the flow to have two options as under:

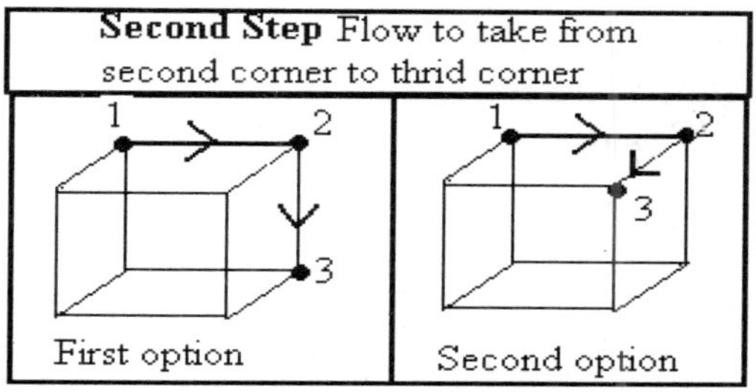

Step3. Either of the options, while reaching third corner, shall be again confronted with a pair of options. The same be followed individually. And, at next step as well there to follow a pair of options, but with that, the rest of the flow is to have only single flow option to connect remaining all the corner points.

5. It be taken, as an exercise to workout all the options for the original first orientation flow from the first corner.
6. Further it also be taken as an exercise to workout the flow options for the second and third flow orientations from the first corner.
7. Still further, it also be taken as an exercise to workout the similar flow options for all the eight corners of the cube.
8. Still further, as the flow orientation takes from first corner to the 8th corner in seven steps availing

coordination arrangement of seven edges in a sequence, and as each such flow may be taken in reverse orientation, as starting with the 8th corner and reaching up till the first corner, as such the question would stand posed for an answer as to how many pairs of such orientations are accepted by the set up of the cube. Before actual chase and working out of all the orientation flow coordination of all the 8th corner at a time, one may attempt the poser as to, if the total pairs of orientations taking from first corner to eighth corner, and in reverse orientation from eighth corner to first corner, to be precisely $8 \times 3 = 24$.

9. One may pause and have a fresh look at the set up of the cube and to chase and comprehend on the format of the flow orientation connecting a pair of corner points in terms of a connecting edge, as that the edge, as a line is a flow of a moving corner / point.
10. One may attempt definition of a line as a flow track of a moving point.
11. One may attempt chase path for a flow track along the format of a line in a pair of ways parallel to the pair of the orientations of a line.
12. One may attempt to define a direction for the other end corner point in terms of the orientation of the edge leading from first corner point to the second corner point.

■

5

EQUIDISTANT FROM THE CENTRE

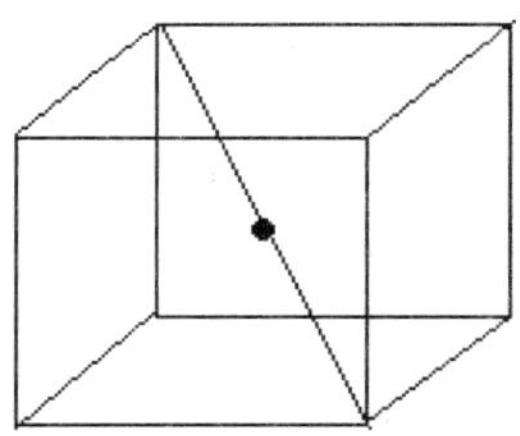

1. One shall have a fresh look at the set up of the cube and see that all the corner points of the cube are equidistant from the centre of the cube.
2. It would be a good exercise to think of an exercise to measure and conclude as that the corner points are equidistant from centre of the cube.
3. As a reverse exercise, the above property of the set up of the cube may be availed to fix a centre of the cube.
4. In the light of the above property, and attempt may be made to define a centre of the cube as a point inside the cube as being equidistant from all the corner points of the cube.
5. It further would be a good exercise to demonstrate

that there is no other point of the cube which is equidistant from all the corner points.

6. This demonstration may be worked out with the help of a soap cake and a needle. 'One may, with help of a big needle reach from one corner of an edge to centre of the cube and ahead to another corner of the cube. This way one can measure the distance between such pair of corner points. This exercise may be repeated beginning from all the corner points and see that there would be only four pairs of corners which shall be along the needle passing through the centre. The length of the needle connecting such pairs of corners may be concluded as of double the distance between the centre of the cube and any of the corners of the cube.'
7. For firm comprehension of this feature of the set up of the cube, one may have other demonstrations like measuring the distances between a corner points from inside of a room of cubic dimensions with the help of a bamboo.
8. The glass cubes with small holes at corners and the needle and a thread may be the another facility which when suitably availed is to make it possible to demonstrate equidistant feature of centre and the corner points of the cube.
9. The above facility of a glass cube and the needle and a thread may be availed for demonstration of other features of the set up of the cube by actual measurements of lengths, angles, areas, volume and interconnected relations of corners with centre, edges with centre, surfaces with center and of other points of the cube with centre. ■

6
LEARNING FACILITIES

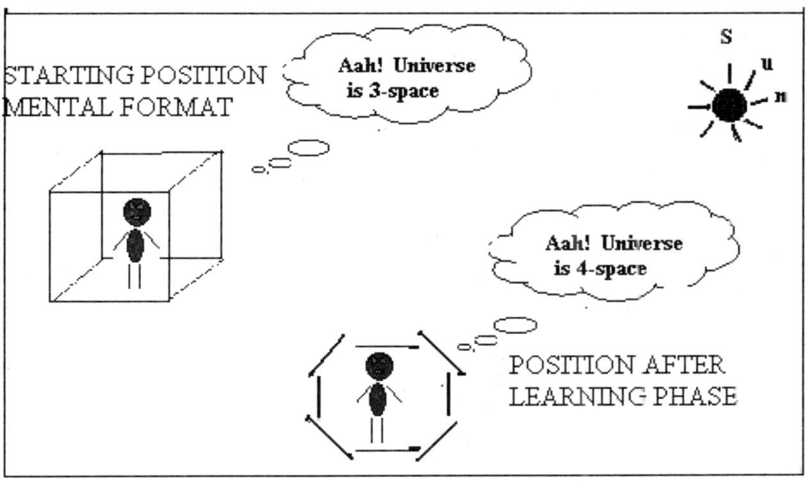

As the learning process is to avail 'cube' as the startwith set up for learning, as such, it is to be the 'first learning facility'. Further as, the mind is seated within the human body, as such it is to be 'the second learning facility'.

This pair of learning facilities shall be helping to approach the 'existing mental format of young minds' for transition and transformation for 'next phase and stage mental format of mature brains'. It is attainment of this transition and transformation which is to be taken as a success of learning of this phase and stage of schooling.

The success range of learning of this phase and stage of schooling is to be taken to be as is the range between the starting position of the mental format and the attained state for the mental format at the end of the learning.

These end points of the range, may better be taken as the *sadhka* itself as the starting point and the guru as the end point. This, in the language of the Discipline of learning, may be described as the start with mental format of the student being as his existence domain being that of a cube and its attainment state being the domain of a hyper cube4.

'Sun', in the scheme of this system, is the source reservoir, and as such the scriptures preserve as a distinguishing feature of the system accepting the range of learning as from 'Earth' to 'Sun' with the first step being the performance of 'Surya Namaskar' / faithful salute to the Sun.

■

7

VEDIC MATHEMATICS SKILLS

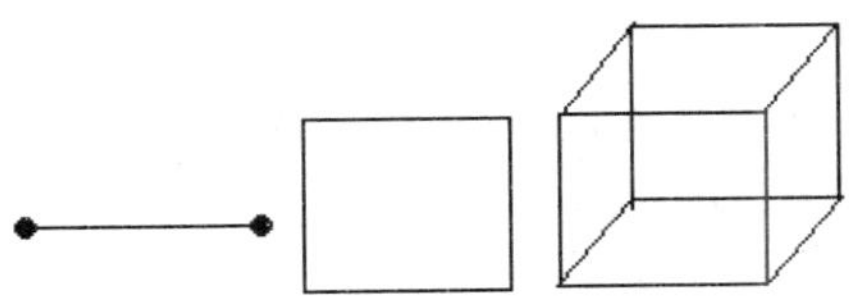

Interval Square Cube

$(X^1 + 2X^0)^N$, N =1, 2, 3

1. Let us look at the set up of an interval and see that this set up consists of the geometric entities (1) the length (X^1) and (2) the boundary / end points $(2\ X^0)$.
2. These entities of the set up of an interval accept combined expression as: $(X^1 + 2X^0)^1$

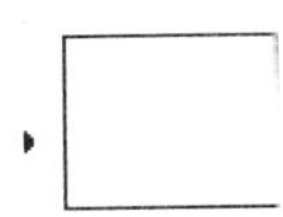

Square

3. Now, let us look at the set up of a square and see that this set up consists of the geometric entities (1) the area (X^2), (2) the boundary lines $(4\ X^1)$ and (3) the corner points $(4\ X^0)$.

4. These entities of the set up of a square accept combined expression as: $(X^1 + 2X^0)^2 = X^2 + 4X^1 + 4X^0$
5. Now, let us look at the set up of a cube and see that this set up consists of the geometric entities (1) the volume (X^2), (2) the boundary surfaces $(6\ X^2)$, (3) the edges $(12\ X^1)$ and (4) corner points $(8\ X^0)$.

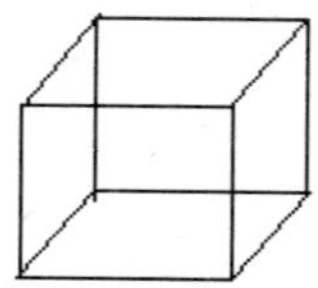

Cube

6. These entities of the set up of a cube accept combined expression as:

$$(X^1 + 2X^0)^3 = X^3 + 6X^2 + 12X^1 + 8X^0$$

7. The above three expressions for the entities of the set ups of interval, square and cube together accept common expression as follows.

$$(X^1 + 2X^0)^N,\ N = 1, 2, 3$$

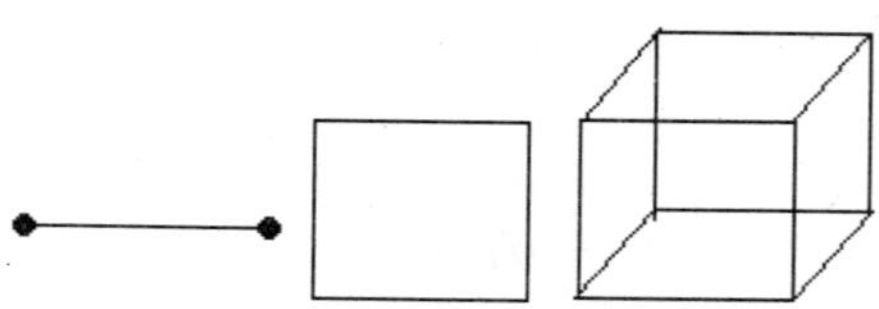

Interval Square Cube

8. One may comprehend as to how the single algebraic expression $(X^1 + 2X^0)^N$, N =1, 2, 3) is successfully expressing about the geometric entities of set ups of all the three geometric bodies

of three different spaces namely interval as bodies of 1-space, square as body of 2-space and cube as a body of 3-space.

9. This as such demonstrates the power of algebra as logic availing first principles simplifies as takes even to the basis of basics.

10. Now, let us have a fresh look at the common formulation

$$(X^1 + 2X^0)^N, \ N = 1, 2, 3$$

Firstly this formulation can be further generlised, by extending its range by permitting and to take further values viz. , N=0, 1, 2, 3, 4, 5---.

Let us see, where it leads to ?

For n=0, $(X^1 + 2X^0)^0 = 1$, a 'point' set up of 'zero space'

As such, the range 'interval, square and cube' accepts 'point' within a range.

For n=4, it leads to $(X^1 + 2X^0)^4 = x^4 + 8x^4 +$ --, depicting feature of hyper cube-4 as a domain within a solid boundary of eight components.

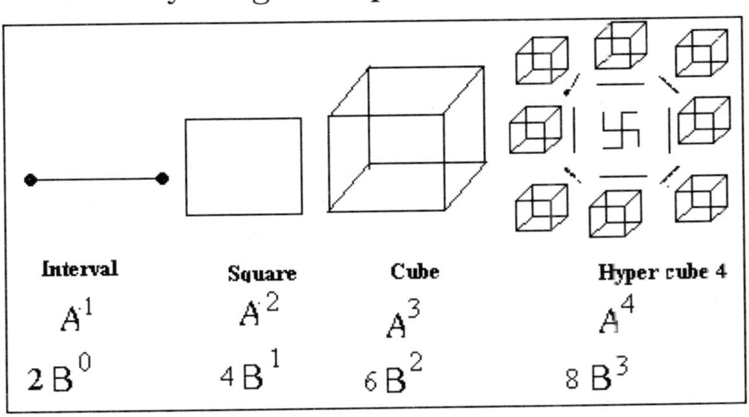

8

PLACE VALUE EXPRESSIONS TO ALGEBRAIC EXPRESSIONS

1. Let us have symbol for 10 as X.
2. The values of powers of 10 may be expressed as powers of X.
3. As such $(10)^0 = 1 = X^0$.
4. Further as that

   $(10)^1 = 10 = X^1$

   $(10)^2 = 100 = X^2$

   $(10)^3 = 1000 = X^3$

   ———————————

5. As such the numerals one to nine, parallel to their '1' as place value, would accept place value and algebraic expressions as follows:

   Place value expressions | Algebraic expressions

10^2	10^1	10^0	X^2	X^1	X^0
		1			1

$1 = 1 \times 10^0$ $\quad$ $1 = 1 \times X^0$

$2 = 2 \times 10^0$ $\quad$ $2 = 2 \times X^0$

$3 = 3 \times 10^0$	$3 = 3 \times X^0$
$4 = 4 \times 10^0$	$4 = 4 \times X^0$
$5 = 5 \times 10^0$	$5 = 5 \times X^0$
$6 = 6 \times 10^0$	$6 = 6 \times X^0$
$7 = 7 \times 10^0$	$7 = 7 \times X^0$
$8 = 8 \times 10^0$	$8 = 8 \times X^0$
$9 = 9 \times 10^0$	$9 = 9 \times X^0$

6. The place value expression for two digit numbers namely, 19, 28, 37, 46, 55, 64, 73, 82 and 91 shall be having corresponding algebraic expression as follows:

Place value expressions			Algebraic expressions		
10^2	10^1	10^0	X^2	X^1	X^0
	1	9		1	9
	2	8		2	8
	3	7		3	7
	4	6		4	6
	5	5		5	5
	6	4		6	4
	7	3		7	3
	8	2		8	2
	9	1		9	1

$19 = 1 \times 10 + 9$	$19 = 1 \times X + 9$
$28 = 2 \times 10 + 8$	$28 = 2 \times X + 8$
$37 = 3 \times 10 + 7$	$37 = 3 \times X + 7$
$46 = 4 \times 10 + 6$	$46 = 4 \times X + 6$
$55 = 5 \times 10 + 5$	$55 = 5 \times X + 5$
$64 = 6 \times 10 + 4$	$64 = 6 \times X + 4$

$73 = 7 \times 10 + 3$	$73 = 7 \times X + 3$
$82 = 8 \times 10 + 2$	$82 = 8 \times X + 2$
$91 = 9 \times 10 + 1$	$91 = 9 \times X + 1$

7. Like that every number of ten place value system can be expressed in algebraic form.
8. In fact any number expressed in any place value system can be re-expressed in algebraic form.
9. Lesson-1 takes us to the conclusion as that the geometric entities constituting different set ups of interval, square and cube as bodies of 1, 2 and 3-space respectively, can be expressed in terms of general algebraic form. The conclusion of this lesson as that numbers of all number system as well can be expressed in algebraic forms.
10. Once numbers and bodies accept expressions in algebraic forms, the different branches of mathematics get unified.

■

9
ONE COUNT

A WORD WITH THE TEACHERS

Being a start with position for initiation of the learning for the middle school levels students, and the learning being aimed to be on first principles availing basis of basics of mathematics as chased by the ancient wisdom discipline of Vedic mathematics, as such, responsibility lies upon the teachers to gently expose the young minds to the delicate features of their this phase and stage of learning. This lesson content of this section, is in fact meant for sharing with teachers, the aspects of the 'path of learning through which the learners have to be taken through' as along this path they have to carry the young minds along with them. As such, the lesson is being numbered as lesson 00, as, the same is being shared with the teachers.

1. The basics of numbers as per the ancient wisdom is centred around the features of 'one', as 'one count'.
2. 'One', otherwise is as a whole some concept, as much as that 'one' is the whole Brahmand / universe, and simultaneously, it is, its smallest

possible constituent, and even that which is devoid of anything, and as such is conceptualized as 'zero', as well is 'one'; 'zero' as 'one count'.

3. This way, the basic of all basics, to be designated as 'basis of basics', is 'one', being 'sole'; 'whole', and everything 'from sole to whole'.
4. As such, the young minds, lead be exposed to the learning of 'numbers', and all that is formally taken as 'Arithmetic', in its specifics as well as in its generality of the discipline of features of 'artifices of numbers' of all possible coordination and arrangements of their organization formats, by focusing upon the features of 'one', as a beginning as well as the end and whole range in between of the Discipline of 'numbers'.
5. This being so, when the young minds are to be exposed and initiated into the 'Domain of numbers', for its learning on its first principles in the light of the 'basis of basics of the Discipline', it is not only at the initial beginning stage, and also not at the final stage but throughout the learning range, at every possible opportunity, the attention of the experience need be consciously focus for in fresh look at the 'features of the whole some 'one'.
6. The young minds are to be very gently taken through transition of 'one' as a count, to the whole range of counting numbers.
7. It would be a real attainment for the teacher when he is satisfied as that the student has really comprehended the counts.
8. Once the 'counting' has come within the comprehension range of the young minds, the rest

is just a sequential process of the learner's intelligence field, and the role of the teacher, as such is to remain that of a guide equipped with the milestone information of the knowledge path along which both are to the traveller companions.

9. This much, being the importance of comprehension of the 'counts', one shall perfect one's counting skill, with the help of the set up of a 'CUBE' as well as, with the help of other expressions along geometric formats, may it be a tree, or a forest or the Earth.
10. The counting skill helps chase the order which even binds the affine setting, as ultimately, even the affine setting as well is to avail a geometric format, and geometric format is to accept counts in their features of artifices of their values running parallel to the dimensional set ups as basis of the geometric formats of affine settings.
11. This being so the basics of numbers with focus upon 'count one' deserve to be chase on their first principles as or the basis of the functional rules of 'Ganita Sutras'.
12. With this the focus of learning the numbers system, at a step ahead of comprehension of 'count one' shifts to 'Ganita Sutras', whose very first Sutra, picks up the thread, straight from the artifices of 'count one', and takes a leap forward of synthesizing a stitching thread parallel to the order of the affine state by taking 'from a given count to the next count'.
13. This, this way synthesized stitching thread for ordering parallel to the format of affine state

setting, attains the whole range of counts of setting along so synthesize stitching thread and thereby all the counts stand synthesize as of a setting stitched with such synthetic thread, as a result of which, the counts emerge to be of the sequential order parallel to the sequential order of the features of dimensional formats originally availed for affine settings.

14. As such the formal teaching and learning of the discipline of numbers stands settled to be a chased in terms of the functional rule of Ganita Sutras, beginning with the ordering of counts as a sequenced stitched by the synthetic thread woven in terms of the functional rule of Ganita Sutra-1, which reads and sounds as under.

GANITA SUTRA-1

एकाधिकेन पूर्वेण

Ekadhiken Purvena

One more than before

FORMAT OUTLINE OF WORKING RULE

15. The rule 'one more than before' synthesizes the stitching thread, as well as sequences 'counts'

which further workout parallel feature of dimensional formats beneath the affine settings of manifested creations fulfilling the Brahmand / Universe, for whose order are the applied values of numbers.

16. This being the order, and so being the features of the artifices of numbers, and all being interwoven and connected, as such the young minds at their initial exposure and initiatives for comprehending this order, deserve to be gently taken through the basics with their basis in terms of first principles by availing the functional rules of Ganita Sutras. And the teachers attending this for the young minds shall be discharging their solemn responsibilities whose performance is going to be its own reward.

■

10

GANITA SUTRA-1

एकाधिकेन पूर्वेण

Ekadhiken Purvena

One more than before

SEQUENTIAL PROGRESSION RANGE

1	2	3	4	5	6	7	8	9	10
ए	क्	आ	ध्	इ	क्	ए	न्	अ	प्
11	12	13	14	15	16				
ऊ	र्	व्	ए	ण्	अ				

1. The first formal step for learning of mathematics is the functional rule of Ganita Sutra-1 which takes

from a given 'count' to the next count as being 'one more than before'.

2. This functional rule is the 'basic of all basics', as that, it is the basic feature in terms of which the 'counts' accept ordering. It is 'basis of the basics', as it also, brings 'one' itself as well within this very ordering rule as 'one is one more than zero' which is occurring before one'. It also leads to the first principle for chasing the functional rule itself by availing the basis of the basics feature as the principle of mathematical induction in terms of which it would be proved as that in terms of the functional rule of Ganita Sutra-1, the entire range of counts would get ordered to be chased and attained for their pure and applied values.
3. The principle of mathematical induction is that when the particular feature value is true for a given count and the same also holds for the following count then the feature value is to be taken as being true for the whole range of counts.
4. Here the functional rule 'one more than before', as it is stated in its generality, when the same is holding true for 'one' as it being 'one more than zero' which is occurring before 'one', and under the rule, it is to be true for the following count, that is for 'two' being 'one more than before', so the same becomes true for reaching 'two consecutive steps' and as the same is to sequentially follow, so it is, in fact, amounts to proving the truth of the principle of mathematical induction itself.
5. This simultaneous unfolding of functional rule of

Ganita Sutra-1 'one more than before', of all the three features, namely 'the basic feature', 'the basis of the basic feature' and 'the first principle' to reach at the basis and basics, makes the functional rule and hence the Ganita Sutra-1 being the par-excellence.

6. This is, in fact, the first glimpse of this par-excellence of Ganita Sutra-1. The studies reveal that the Ganita Sutra-1is the source Sutra, as much as that it itself in terms of its own functional rule not only unfolds its organization format coordinated arranged as a structured set up of sixteen steps manifesting as composition of sixteen letters but also parallel to these 16 steps of the composition of the text of Ganita Sutra-1 are the sixteen steps of the organization format of whole range of text of Ganita Sutras one to sixteen, as well as of Ganita Upsutras 1 to 13. One glimpse of this richness of Ganita Sutras of Ganita Sutra-1, and as a consequential thereof all the Ganita Sutras may be had from the Book of the Author.
7. The functional rule 'one more than before' together with the rule of Subsutra-1, as 'to follow the proportionate symmetry of the form within frame', supplies the basis of all 'basic operations' of the family of 'numbers system' structured on the 'basis of the basics' of 'count one'.
8. The functional rule 'one more than before', state a way becomes the basis of the basics of 'addition operation'. The transition from 'addition to subtraction' as well follows just as a reversal of the steps of operation of the functional rule of

Ganita Sutra-1, as in 'addition' it is to reach the following count as 'one count more', and in subtraction operation, it only would mean to reach the previous count as 'as of values of one count less'. Moreover, the functional rule of Ganita Sutra-14 being 'one less than before', is particular unfolding of Ganita Sutra-1 as the 'source sutra'.

9. The repeated application of the functional rule of Ganita Sutra-1 smoothly attains a transition from 'addition operation' to 'multiplication operation'.
10. Just as multiplication operation, emerges to be, nothing but 'repeated addition', like that ' the repeated subtraction' smoothly transits into 'division operation'.
11. Further like addition and subtraction being interconnected through reversal of steps, the multiplication and division as well are similarly interconnected through reversal of steps.
12. Still further, the repeated multiplication, smoothly transits into exponents, and repeated division transits into decimal conversions.
13. The functional rule of Ganita Sutra-1 is of universal application, as that it is not restricted to number system of any place value system, though Ganita Sutra-1 makes it applicable to the specific formats of any given place value system and Ganita Sutra-2 makes it a favoured choice of ten place value system, for the distinctive features of artifices of numbers being of characteristics which run parallel to the reflection and other features of the carriers within rays of the Sun.
14. As such the first focus while learning basic

operations of number system is to be upon the fundamental unity of all the operations as being of basics features of addition operation as straight manifestation of the functional rule of ordering principle of Ganita Sutra-1.

15. Therefore, perfection of learning of the operation number system, full comprehension focus is to be upon the functional rule of the ordering principle of Ganita Sutra-1 simultaneously unfolding, the basics, feature of basics, as well as the first principles for application thereof manifesting as the first operation of numbers known as 'addition operation'.

16. The comprehension of the basics, basis of basics, the first principles, all together manifesting as addition operation, and the same unfolding as subtraction, multiplication, division, powers / exponents, decimals, and all that, the permutation and combination rules, the probability rules, the binomials and polynomials, the dimensional spaces and dimensional bodies features, everything is all to follow in one 'go'.

17. Needless to add that the teachers are vasted with the heavy responsibility to help the young minds to have a smooth transition for them for one step to the next, and from one operation to another.

■

11

ALL OPERATIONS ARE UNIFIED

1. Par-excellence feature of the number system with focus upon the basic count, that is 'one', is that the 'operations', as the 'synthesis' following the 'functional rule' of Ganita Sutra-1 as 'stitching-thread' running through all the 'counts', in the process, inherently and silently 'unifies' them all.
2. But, this state of 'ordering principle of Ganita Sutra-1' manifesting as 'stitching-thread, and all its unified operations', because of the unified characteristics of its operations, though, makes it a selfsustained domain-field, but never the less, is compactified another 'state' of the ordering principle, for whose transition and transformation, in the smooth transcendence step is to be there to 'transcend' from 'count to pick for a pack'.
3. This transition through transcendence from 'counts' to 'picks' for packs, makes original 'stitching thread state of counts' as a particular case of 'picks for a pack along a stitching thread spread format, which may be designated as a 'line', and 'this line' is not necessarily to be a state line, as the stitching

thread placement figure is to be of form, frame and format, as well as of order of the geometric organization of the body of the space to be availed.

4. The shift to this new attained state for the 'counts' transforming as 'picks', deserves to be chased very smoothly as here essentially, and that to, silently, as inherent feature of newly attained state, there is going to be a shift for the very 'stitching spread' which brings into a change for 'operational rule of ordering' from that of 'a line' to 'an angle', which is to focus with intensity upon 'the space spread' in between the pair of 'lines', and this 'whole transformation' is to be for a 'line' to be a 'strip'.
5. The feature of 'count', as 'whole' and 'a single' ordering step in terms of 'stitching thread line', is carried forward for the next state of 'count' as 'a pick', but, the 'pick' being 'whole' is to be in 'linear ordering sense' is to depict the previous state of stitching thread line as a specific case for the new state. The new state has its 'new wholeness' of its own, as of 'of strip ordering format' and the same makes a much difference for its operations, and the same when viewed from the stitching thread line format, it is to yield a spectrum whose comprehension, if not as a bend, is to make a mental cross over situation.
6. As such, heavy responsibility lies upon the teachers towards young minds being initiated for this exposure for a shift from the original 'stitching thread line' processing format to 'the strip domain angle' This, is to be taken as a challenge of the order of transcending from the old mental format

of 'stitching thread line processing' to a new mental format of the order and feature of 'strip domain angle' for processing rules for the operations to be carried with 'picks being provided a coordinated flow within strips domains-fields'.

7. This transition for a shift from that of 'a stitching thread line format' to 'a strip domain coordinated flow within an angle space spread', for its chase, may be provided a geometric formats which would facilitate comprehension of a single stream flow and the spread for multiple streams flow.
8. The 'pick' as a 'count' along a 'stitching thread line' as of a format of one pint flow track, and on the other hand 'the pick' as of a 'strip domain spreads within an angle' is to be of a format of 'distinct generic' as is inherently to be there with a shift from 'line' as of a length as a measuring unit, to that of 'surface in between the pairs of arms of an angle' which is to accept 'area' as a measuring unit.
9. 'Count as a pick of stitching thread line format for a single point flow track' makes the count and hence the pick as well as of a 'whole unit', while 'the pick of strip domain surface within arms of an angle' may be of 'whole unit of an area' but the same, at the same time may as well is to be of a pair of distinct features due to the availability of a pair of arms containing the surface in between as a flow spread. Illustratively, one area unit, may have a split as a pair of flows, as '$2 \times ½ = 1^2 = 1$'with a flow along one of the arms of an angle to be of 'two' as a unit being a 'stitching thread line unit',

and simultaneously the second limb of flow along the second arm of the angle is to be of 'half unit value of stitching thread line unit'. It, contrary to the counts of original states, makes 'picks' to be of splitting features, which being no more there in the original state counts of stitching thread line format.

10. This distinctiveness of the second state of counts being picks of split features makes the operations of this state of counts as to be of distinct features than that of the original first state counts accepting 'count' being 'single' / 'whole'.

11. The wholeness-splitting feature deserves to be fully chased and the same is not to be restricted only up till the flow within the space spread between a pair of arms (of stitching thread line format) of an angle. The arms may be of 'strip flow format'. Then there would be solid arms formats, and the same as well may not be the angels only of a pair or arms as the solid angles may be of three solid arms, and so on. Ahead, there would be of hyper solid formats.

12. The shift from the flow within a space spread contained by a pair of linear arms to the space spreads within a pair of spatial format arms, and of solid and hyper solid formats ahead, deserves to be chased smoothly and sequentially for their comprehension to simultaneously attain in the process the perfection of intelligence as well.

■

12

PICKS OF FULL FEATURE AND OF HALF FEATURE

1. The Creation's secret of picks of different splits is shared with us by the 'cube, our best friend' from the 'World of Creation'.
2. Let us have a fresh look at the affine set up of the cube and see that a flow stream from any corner to the other corner connected through an edge, on reaching to the other end corner brings a phase and stage for the flow to split into a pair of options available their along the other pair of edges flowing out from that corner.

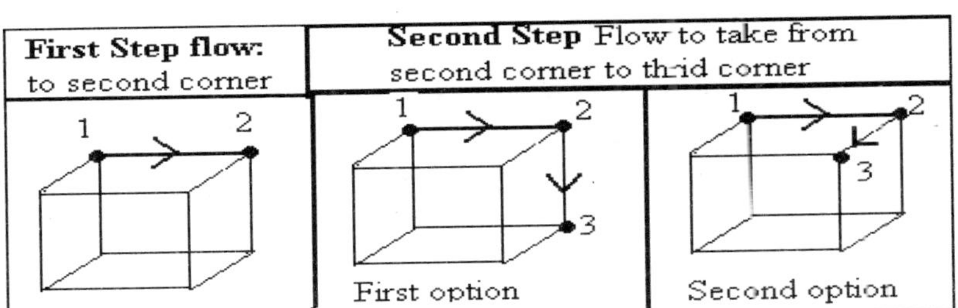

3. Let us chase the flow, a step ahead, and would see that, there as well a split pair of options in either of the second step options, would be

available for the flow as at third step, being of double pair of options as such:

Third Step Flow at third corner as well to be of a pair of options		**Third Step** Flow at third corner as well to be of a pair of options	
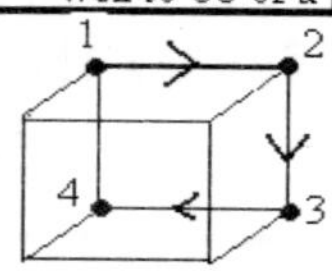 First option	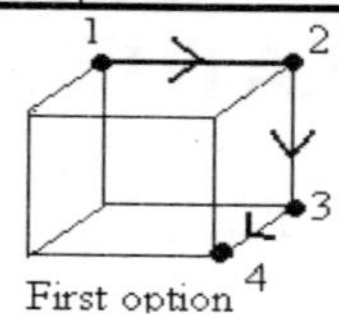 First option	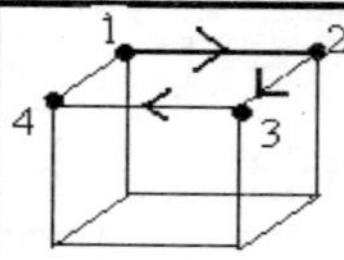 Second option	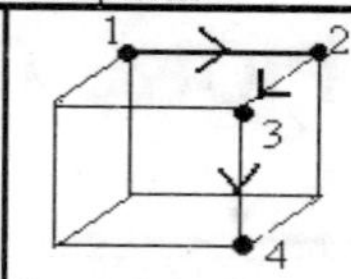 Second option

4. The steps ahead for the double pair options flow up till fourth corner, when to be chased ahead, with a restriction not to come back to the corners already connected, it shall be a good exercise to reach up till the last, that is, the eighth corner of the cube.
5. One such split flow stream through first to the eighth corner of the cube would be as:

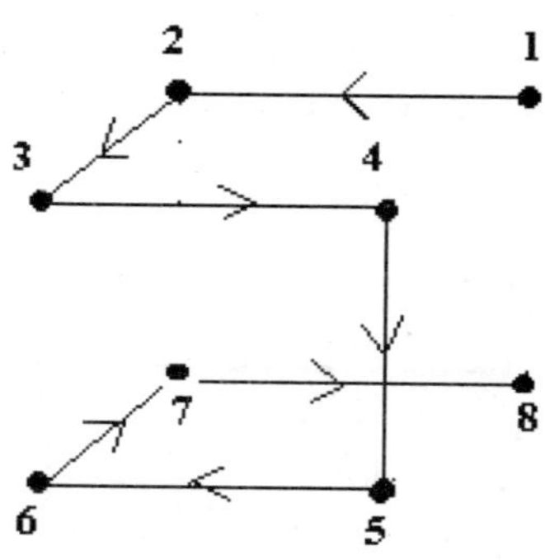

6. It would be interesting to note as that, the reverse of this flow, again would be one such split flow sequence, beginning with the eighth corner as the first one and in that sequence the first corner would be the eighth one.

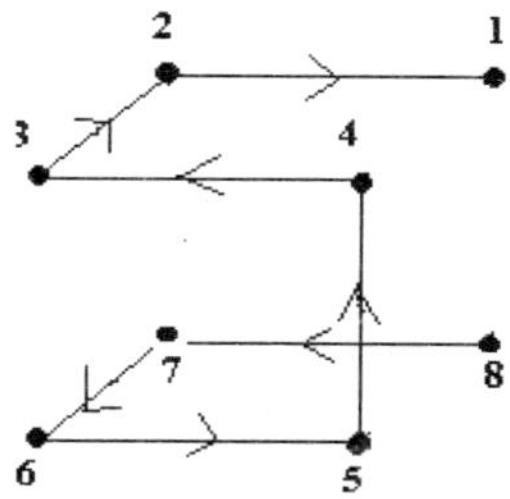

7. One may pair these flow streams as a pair of flow streams of opposite orientations.
8. This makes a pair of flow stream of opposite orientations of seven steps long ranges.
9. One may have a fresh look at these flow streams ranges and observe as to how the flow orientation for the first three steps and of last three steps of these ranges are of opposite orientations availing spatial formats. It is this opposite orientations synthesis through the middle, that is fourth step of the ranges, which makes each flow stream being of such a feature which permits split into a pair of spatial format opposite orientations, and hence the distinctive feature of synthesis of pair of full seven steps long streams flows of opposite orientations being inherently getting inbuilt into individual streams as to be of a pair of halves as of opposite orientation.
10. It is this feature characteristics which coordinates a pair of orientations at each of the seven steps of the flow stream range, which makes it to be of distinctiveness as of 'Sakalana Vyakalna Bhyam / Simultaneous addition and subtraction format'.
11. The first orientation flow as to be track of interval

as a body of +1 space, and the second orientation flow to be the track of interval as body of (-1) space, and thereby, there being, ensured distinctiveness, for both flow channels, because of the intervening '0-space', as of un-manifest but definite separating channel at their middle.

12. As such 'one to one correspondence' rule for counts of original state number system, would follow for both (+1)-space format flow as well as for (-1)space flow but the same to be so only because of third un-manifest flow format for the interveningt 0-space format at the middle.
13. This emergence of its own of third un-manifest at middle of the pair of flow paths formats of split flow for the stitching thread line format into a strip domain, deserves to be availed while chasing the operations for this second state of number systems of counts as picks permitting split paths for the flow.
14. It would be the, phase and stage of the chase which shall be providing transition from applied value of 'one to one correspondence' of the full feature counts system of numbers to the 'half way ride along the flow carriers' of the second state number systems.
15. The young minds deserve to be very gently helped for smooth comprehension of transition from 'one to one correspondence' attainment under the 'count units' with focus upon 'count one' being of a full features, with no option for its split, to the 'half way ride along flow carriers' of the second state 'counts picks' of characteristics permitting split for the flows.

16. This 'half way ride along flow carriers' may for facility of comprehension be illustrated from the different but parallel situation of 'the rider alighting the horse half way the journey'.

■

13

FIRST TO CONNECT THE FOUR CORNERS OF THE SURFACE

1. The flow stream which first connects the four corners of the given surface, as three steps of flow, at its fourth step is left with the full flow feature of single option at the fourth corner.

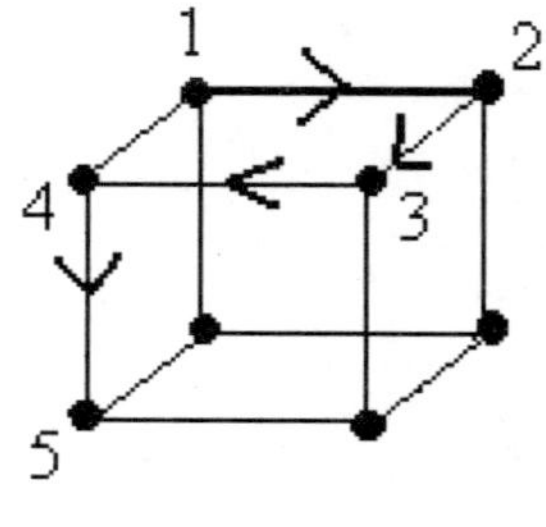

2. This, as such, makes it evident, as that, simultaneously three such flow stream would be available from each of the corner points respectively within three planes meeting at the said corner points. These as such make out a coverage of nine edges (all the four edges for the first plane, three edges for the second plane and two edges for the third plane meeting at the corner).

3. The distinctive feature of the corner points setting of the cube is that precisely three edges and three surface planes meet at every corner point.
4. This, as such, may help comprehend the coordination of arrangement for all the nine vowels by Maheshwara Sutras, as coordination of flow streams. The first three vowels are coordinated constituting an arrangement of three steps, as are to be there a three streams flow from each corner point. The remaining six vowels, as three pairs, are three distinct phases and stages of coordination as Maheshwara Sutras –2, 3 and 4 respectively, as to emerge three distinct phases and stages of flow stream which at their second, third and fourth step are to be of two options each.
5. The same may be chased with the help of following depiction as first step of coordination arrangement of first three vowels as Maheshwara Sutra-1 and parallel to it be the coordination at corner for three fold split of a flow stream.

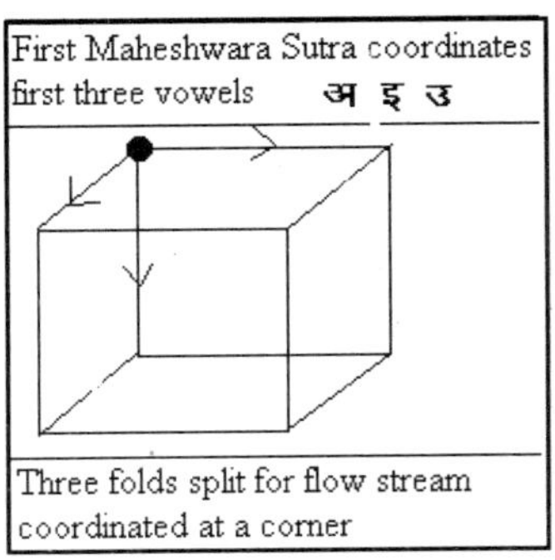

6. As second step, it may be chased as two options at second corner, followed by two option at third corner

and finally further two options at the fourth corner. Then ahead there would be a smooth unified single stream flow till the eighth corner. The three steps of double options and parallel to it the coordination of remaining six vowels as three pairs, in a sequence may be depicted for the each flow stream from the first corner as follows:

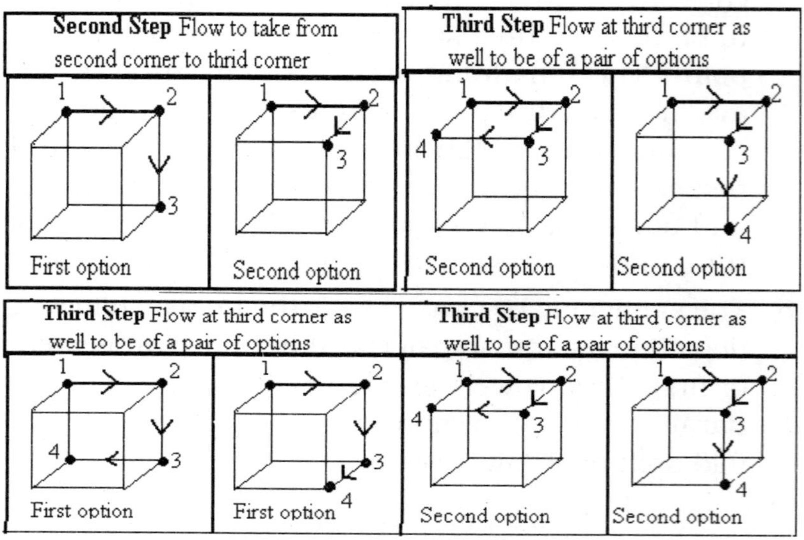

7. The subsequent availability of pair of options as fourth step flow of the flow range, may be chased by depiction as an exercise.
8. The flow charts depictions for other flow streams from corner-1 may be attempted to have full comprehension of the settings of the flow stream ranges.
9. Likewise the flow stream ranges depictions for starting from the other corner points as well may be chased.

10. It may be a good opportunity for self-evaluation of one's comprehension of the flow stream ranges of the settings of the affine setting of the cube by having full length expression of ones comprehension of this feature.

■

14

COUNTS TRANSCEND ALL FEATURES

1. 'Count' transcends all feature of an entity and takes the whole of entity as count 'one'.
2. The entity features of 'cube', as corners, edges and surfaces are dissolved and wholeness of 'corners' or of 'edges or surfaces' earns single count for each of them.
3. All the eight corner points, twelve edges and six surfaces count as 'count twenty six'.
4. 'The counting' is a fine art. The counting skill is the basic mathematical skill. The first mathematical skill to be learnt is of 'counting'.
5. The skill perfection leads to the comprehension of the ordering rule of Ganita Sutra-1 एकाधिकेन पूर्वेण Ekadhiken Purvena / One more than before.
6. The setting of counting numbers as a sequence along stitching thread line of the functional rule of Ganita Sutra-1 earns the designation for such line being the 'number line'.
7. The 'number line' may not necessarily be a state line, and in fact it may be of any setting format,

and the same, of its infinitely many settings, prominently of facility expression may be a straight line, circumference of a circle, the curve along the surface of a sphere, or along or within any solid, or even along any of the hyper solids.

8. Of these the formats for the 'numbers lines', the formats of a straight line, circumference and of a section of a cone.
9. The straight line settings of 'counts' is of popular comprehension, and the 'nine points setting' along the circle as well is of equal popularity, and third in the row of this prominence is of a number cone.

Different formats for number value line		
Straight line	Circumference	Section of a cone
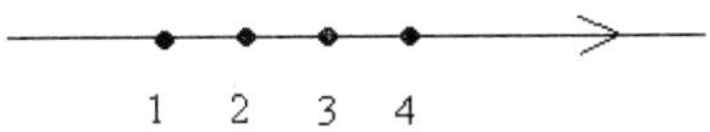	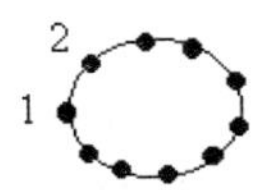	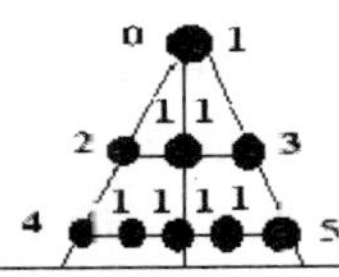

10. The number value line of section of a cone format above, further leads to a feature of split of counts as even counts and odd counts being distinct streams of counts.
11. The 'even counts stream' also focus another feature of counts with 'two as one count'.
12. This counts line of even counts of number cone is also designated as a line of 'table of two'.
13. This, as such, silently takes us to the 'rule of proportionately' of Ganita Upsutra-1 आनुरूप्येण / ANURUPYENA / 'to follow the proportionate symmetry of form within a frame' / proportionately.
14. 'Two as one / 2 as 1' is of proportionate symmetry

to 'one as one / 1 as 1'.

15. Likewise can be chased '3 as 1', '4 as 1' and so on.
16. This, this way, firstly make the Ganita Upsutra-1 as a particular case of Ganita Sutra-1.
17. Secondly this makes the table of two as wellbeing a particular case of 'the counts' and the number line in a particular case emerges to be the line of 'table of two'.
18. The pair of arms of section of a cone as a number cone being of a split of a stitching thread flow of flow lines of count into a pair of flow lines, under the proportionate symmetry rule shall be sequentially taking us to triple, quadruple and multiple armed number cone.
19. This flow split and the parallel formats of number cones shall be leading us to the tables lines of whole range of counts as single line flow / pair of lines flow / triple lines of flow and so on.

Single flow line Table of 1	Double flow lines Table of 2	Triple flow lines Table of 3

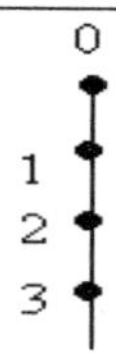

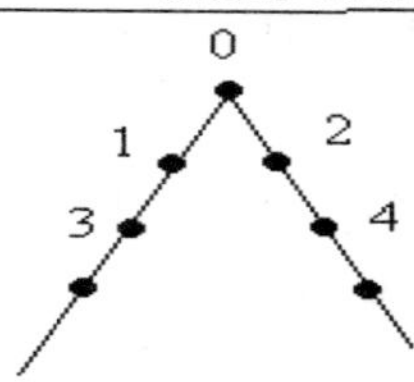

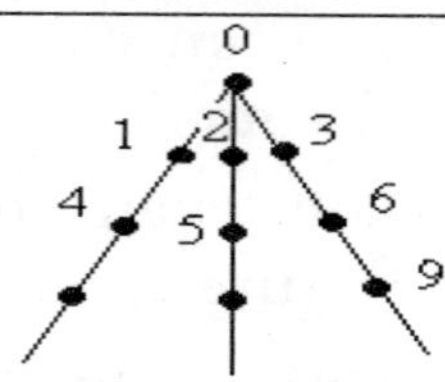

■

SECTION - IV

FEATURE OF SKILLS AND ATTAINMENTS

1

INTERLINKING NOTE

This shall be taking to the fourth phase and stage of present set of series of books of Vedic mathematics basics.

The first three of these focus upon (1) Art of teaching (2) Science of learning and (3) Vedic mathematics Basics.

Present the book-1 which focuses upon the art of teaching and is essentially 'teachers centred'. It is only, after the students are through this phase of acquisition of skills, and have been through the books 2, 3 & 4 about glimpsing of Reality that transcends 3-space and gives vision of 4-space, the students can revert to book-1.

However, to begin with, students may begin with book-2 which is about the science of learning and the same is 'students centred'.

The present phase and stage of learning is of acquiring skills of Vedic Mathematics for Perfection of Intelligence by chasing the features of the set up of a cube.

The focus here is upon following the coordination arrangements of organization format of a set up of a cube. The beginning is had with superimposition of grids along the surfaces of the cube and first chasing the way solid domain manifests and stand contained within the geometric envelope of a cube.

This essentially is the phenomenon of flow of content through the centre of the grid zone and its manifestation as layers of surfaces superimposed upon each other and getting compactified as a domain fold of the content lump.

The skills acquisition runs parallel to the way the streams of content within the domain as it stands contained and framed within the geometric envelope of the cube.

The perfection of intelligence attained in the process becomes capable of chase of streams of content flow with 'counts' as flow steps in terms of fractions of picks of counts as measure units.

With it the mathematics of counts of first count as 'one', sequentially transits and transforms as of 'fractions of picks' of counts for coverage of the flow streams of content lump of 3-space domain within the geometric envelope of the cube.

These skills, not only result into perfection of intelligence as a process of learning but also attained transition for the linear order of the set up of 3-space / cube into spatial order of the set up of 4-space / hyper cube4.

It is the split of the flow into a pair of streams availing artifices '2 and ½' simultaneously, which together workout a spatial unit $1^2 = 1 \times 1 = 2 \times \frac{1}{2}$.

It is this to be simultaneous of by '2 as 1' and also ' ½ as 1' along the pair of axes of a surface, which so chased gives an insight into the process of transition from linear order of a set up of a cube / 3-space into the spatial order of the set up of hyper cube4 / 4-space.

As such, the skills learning sequence is to begin with the comprehension of 'one' itself. In fact, the beginning and the end of this chase of skills range is to be the comprehension of 'one itself'.

It is not only the beginning and end of learning of this range is to be with comprehension of 'one', but the whole range in between as well is to be nothing but the expression of the different features of 'one'.

One shall always look fresh at one's comprehension of 'one' and one would be finding as that one's previous comprehension of 'one' was restricted up till a limited range and that new features are continuing coming within the comprehension range.

Sadhkas shall continue extending their comprehension of the range of features of 'one' and with it go on intensifying their acquisition of skills for perfection of their intelligence and in the process also understanding fully the basics basis of Vedic mathematics on first principles which shall be making them capable of glimpsing the Reality which transcends the set up of 3-space / cube and takes to the Creator's space / 4-space / hyper cube4.

■

2
INTRODUCTORY

1. Vedic systems accept whole range of knowledge as a single discipline.
2. As a result, every Vedic system, may it be mathematics, acquires the features of wholeness of the domain of knowledge.
3. It is because of this feature of wholeness acquired by Vedic mathematics systems, every mathematical branch ultimately accepts the same set of basics.
4. May it be arithmetic, algebra or geometry, all get unified for their basics, basis of basics as well as the set of first principles to be availed to workout the systems.
5. Ganita Sutras, including Ganita Upsutras, emerged to be the single source of the Vedic mathematics systems for the Vedic mathematics basics in its all folds of basics, basis of basics as well as of the set of first principles in terms of which the mathematical system are to be worked out.

6. As such the essential mathematical skills required to be learnt are supplied by the Ganita Sutras, which deserve to be chased as a source for the functional rules of the skills of mathematical operations and for all other pure and applied values of mathematics.
7. The comprehensive view of the prominent mathematical skills may be had by chasing the functional rules of Ganita Sutras (and of Ganita Upsutras) in the sequence and order of the Sutras themselves.
8. The beauty of the Ganita Sutras lies in their organization formats as that while the Ganita Sutras themselves provide the basics, the organization formats of the Sutras provide the basis of those basics.
9. Further the beauty of the Ganita Sutras systems is that the functional rules of the Sutras are selfreferral, as much as that the very functional rule of the Sutra unfolds from within to lead to the basics as well as the basis of the basics of the functional rule itself. This way the first principles for working out the functional rules as well stands supplied by the system itself.
10. Further the Ganita Sutra-1 emerges to be the source Sutra, sequentially unfolding its own composition as well as the organization format from the first step of first letter of the text of the Sutra to the last step of last letter of the Sutra. Not only this. The whole range of the functional domain from Ganita Sutra-1 to Ganita Sutra-16, and also of Ganita Upsutra-1 to Ganita Upsutra-13, as well

sequentially unfolds parallel to the sequential unfoldment of Ganita Sutra-1 itself from the organization format of its first letter to its last, the sixteenth letter.

11. This selfreferral feature of the Ganita Sutra system is inherently inbuilt in the organization format of Ganita Sutra-1 and further this feature of Ganita Sutra-1 follows from the transcendental prefix (*Om*) being accepted by the scripture of Ganita Sutras.
12. The transcendental prefix *Om*, as *'Om iti ek akshar Braham'* leads to the focus upon the transcendental feature *'ek'* that is '*Om* as one', and with it the Vedic mathematical system starts unfolding from within (one itself).
13. With it the sequential chains, and also the ranges of these chains, constituting mathematical systems, and their skills, when to be chased, the beginning of it naturally is to be had with comprehension of 'one'.
14. Therefore the first mathematics skill to be learnt by the students of mathematics is to fully comprehend 'one'. It is with comprehension of one, the minimum of the half of the job may be taken as having already being successfully completed. For many, this success may be 'full'; with their full comprehension of all the features of one. Vedic system accept as that who knows one has nothing more to know. The whole range of knowledge, including mathematics is taken to be completely acquired with complete comprehension, understanding and knowledge of the wholesome sole syllable *'om'* being *'ek ekshar braham'*. ■

3

FULLY COMPREHEND THE FEATURES OF 'ONE'

1. First mathematical skill is to know 'one'.
2. One may accept one's mathematical success parallel to one's comprehension of 'one'.
3. The one who wants to have full mathematical success is to fully comprehend all the features of one.
4. As such the *sadhkas* fulfilled with intensity of urge to fully know and to completely chase the features of 'one' shall go in trans time and again and to glimpse all the features of 'one' as '*ek akshar braham*' lively at the core of transcendental worlds.
5. The Sanskrit formulation 'ek' 'एक' avails sixth vowel, being *shan-mukha* 'ए' / Lord Vishnu, as its first letter and it leads to the first consonant as the second letter, being *chatar-mukhi* 'क्' / Lord Brahma.
6. This is a measuring unit supplied by Creator's space (4-space) for the measuring rod of Vishnu lok (6-space) which is to constitute the core of the transcendental worlds (5-space).
7. As such, this first skill deserves to be acquired

smoothly and sequentially by permitting the transcending mind to glimpse the **inner-folds** of the transcendental worlds and also the **core** of the transcendental worlds.

8. The formulation 'एक', 'ek' 'one' which on its **reverse format** being 'अक्ए', the same to be of the range of *partihara* 'अक्' / expression format supplied by *Maheshwara Sutras* of the range of first five vowels 'अ इ उ ऋ लृ', to be followed by the sixth vowel, in that sequence and order leading through the range of transcendental worlds to its **core** being the seat of Vishnu lok.
9. This way, the knowledge at the base of this skill of comprehension of 'one', is to unfold the whole range of knowledge in one go.
10. That is why, it is only with perfection of this very first skill of knowing 'one' by permitting the transcending mind to glimpse the inner folds of the transcendental worlds as well as the 'affine core tip range' of the transcendental worlds is to mean to attain complete enlightenment and nothing more would remain to be learnt by those who really know 'one' as of its all features being of the order of the Vishnu lok as core of the transcendental worlds.

■

4

INTERLINKING NOTE-2

This shall be the final phase and stage of present set of series of five steps long range being book no. 1 to 5 as five folds of Vedic mathematics basics.

The first four folds focus upon (1) Art of teaching (2) Science of learning (3) Vedic mathematics basics and (4) Skills of Vedic mathematics for perfection of intelligence.

The final fold is the phase and stage of attainment of glimpsing the Reality that transcends the set up of a cube and takes us up till hyper cube4.

The focus here, and as such of the range of teaching and learning of this phase and stage of schooling, is to attain transition from the set up of a **cube** to that of hyper cube4. With this, the focus shifts from cube to that of hyper cube4.

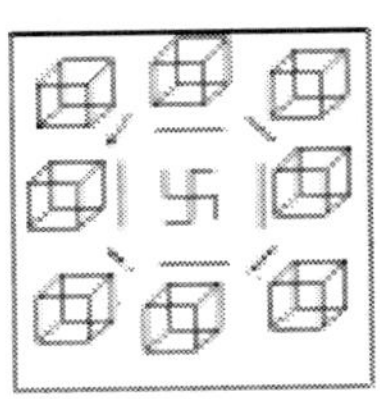

The first lesson of this phase and stage as such comes to be to glimpse as that hyper cube4, as well is accepting common formulation for its domain and boundary along with interval, square and cube as $A^N : 2^N B^{N-1}$, N=1, 2, 3, 4.

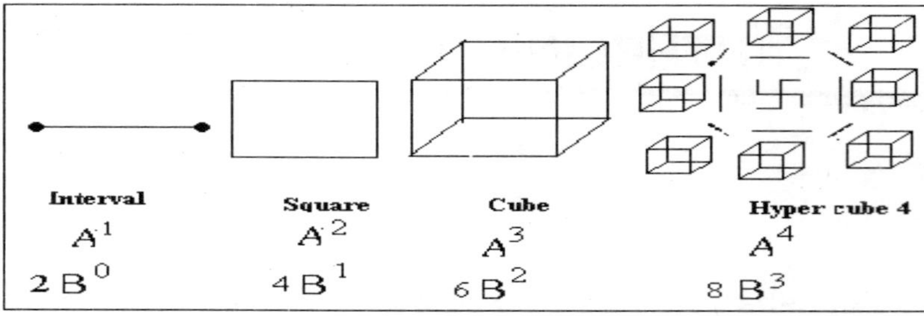

At first step, as book-1 focusing upon art of teaching of Vedic mathematics, we had started from 'the foundation' for the chase of the range of manifestation of solid quantifiers within Creator's space (4-space) beginning from the very stage of their emergence from the transcendental core of the Creator's space.

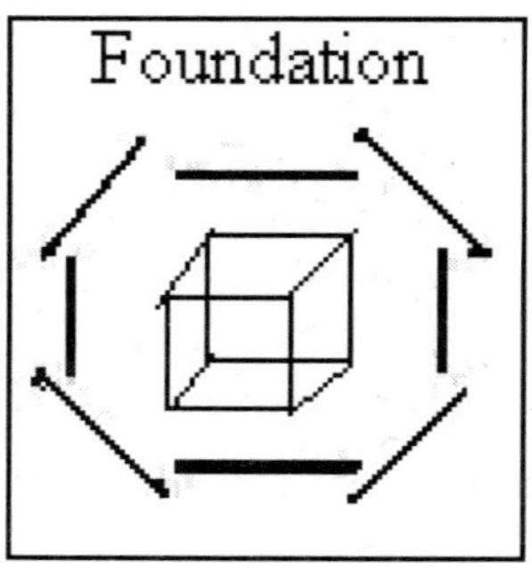

The next, phase and stage has been of book-2 of the science of learning of Vedic mathematics. Its starting point has been that of a cube fixed in its static position on a four legged table.

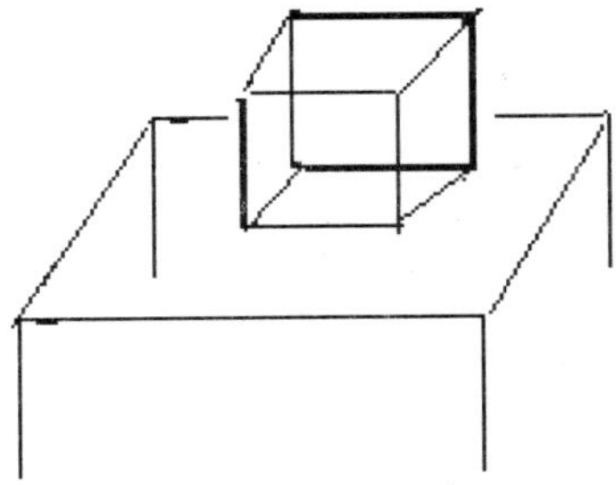

The third and fourth stages has been of learning basics and skills of Vedic mathematics. Book-3 titled basics of Vedic mathematics unifies the set ups of interval, square and cube and beginning for it is had in terms of the common formulation.

$$(X^1+2X^0)^N, N=1, 2, 3.$$

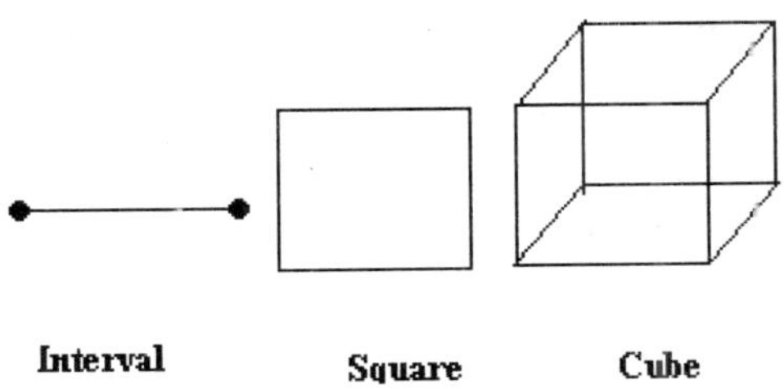

$$(X^1 + 2X^0)^N, N = 1, 2, 3$$

$$(X^1 + 2X^0)^1 = X^1 + 2X^0$$

$$(X^1 + 2X^0)^2 = X^2+4X^1 + 4X^0$$

$$(X^1 + 2X^0)^3 = X^3 +6X^2+12X^1 + 8X^0$$

Book-4 is the phase and stage of acquisition of knowledge for coordination arrangement of organization format of a set up of a cube beginning with a step of superimposition of grids along the surfaces of the cube and chasing the solid domain features.

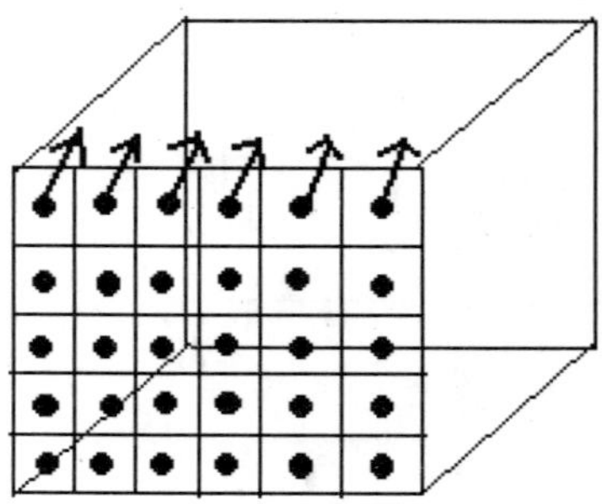

With this, the final stage of transition from the set up of a cube to that of the hyper cube4 shall be putting *sadhkas* realizing their start with mental format as of a situation of being within a cube and after the learning stages having started visualizing the Reality that transcends 3-space set up.

It is this glimpsing feature which is to be taken as the blissful fruit and attainment of the *sadhkas*.

For satisfaction of further urge as to the Reality that transcends the set up of hyper cube4 itself, naturally the *sadhkas* are to continue learning further, for which are all heart full good wishes for all of them.

■

5

FOCUS TO REMAIN CENTRED AROUND CUBE

1. Present series of five volumes of Vedic mathematics basics are being volume wise by phased, as book-1 'The teaching of Vedic mathematics' basics, book-2, 'Vedic mathematics basic', learning, book-3, 'Basics basis on first principles', as the properties of numbers, book-4 'Vedic mathematics skills', as logic of algebra and book-5 'Vedic mathematics projections' of higher dimensional Reality.
2. One way to look at the theme of series is being of five steps of teaching and learning of Vedic mathematics on first principles of basis of basics to be projected for comprehension of higher dimensional Reality as a goal.
3. Teaching of Vedic mathematics of its basics, in itself, is a part of the ancient discipline of Vedic mathematics.
4. The Art and Science of teaching of Vedic mathematics is intimately connected with the process of learning of basics mathematics itself.

5. As such, the Teaching of mathematics, as well as the Learning of Vedic mathematics, despite being intimately connected, are distinct topics of the Discipline and the same deserve to be approached independently and distinctively as distinct arts and sciences in themselves.
6. The basics of the ancient wisdom of Vedic mathematics as well are of features which do not remain restricted only at mathematical results level, but the same transcend and reach at their basis as well, as the first principles in terms of which these can be approached to work out these basics as mathematical attainments.
7. These three fold features of basics as (1) Basics (2) Basis and (3) First principles, manifest as the properties of 'numbers', the 'logic of algebra' and 'geometrical projections' for chase of 'higher dimensional reality'.
8. As such, one way to phase and chase the ancient discipline of Vedic mathematics at the phase and stage of middle school level of instructions, is to be of five steps, to be designated as (1) Teaching (2) Learning (3) Arithmetic (4) Algebra and (5) Geometry.
9. The present volume, as such, is essentially about 'geometry', but the same is of features which inherently are of different characteristics than those of the modern day main stream Geometry which has been restricted only up till linear dimensional 3-space and that too for a very few properties.
10. Vedic geometry, on the other hand, does not

restrict itself only up till 3-space. It transcends much beyond to 'four and higher dimensional reality'.

11. However, the existing hold of present day, 'teaching and learning of geometry', it is taken by conscious decision to remain focused, at this stage only on the phase of transition and transformation from the set up of a cube, as a representative regular body of 3-space to that of the set up of hyper cube4, being the representative regular body of 4-space.
12. As such, essentially, the focus here is to remain centred around the 'cube' but with an aim to have smooth transition to the set up of 'hyper cube4'.

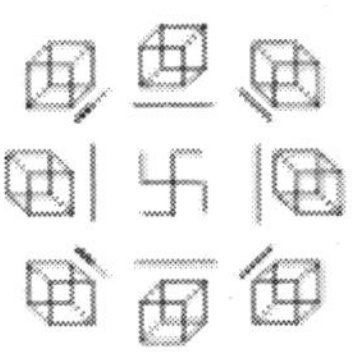

13. The symbolic representation for 'hyper cube4' is being taken here as distinctively depicting solid boundary of hyper cube4 as of eight distinct components, and that to as eight cubes.
14. Hyper cube-4, for its set up, as it is, is in continuity of the set ups of interval, square and cube, as much as that all these four bodies including Hyper cube4 accept a common formulation for their expressions in terms of their 'domains and boundaries' being as follows:

$$\mathbf{A^N : 2^N B^{N-1}, N=1, 2, 3, 4}$$

■

6

DOMAIN BOUNDARY FORMULATION

A^N: 2N B^{N-1}

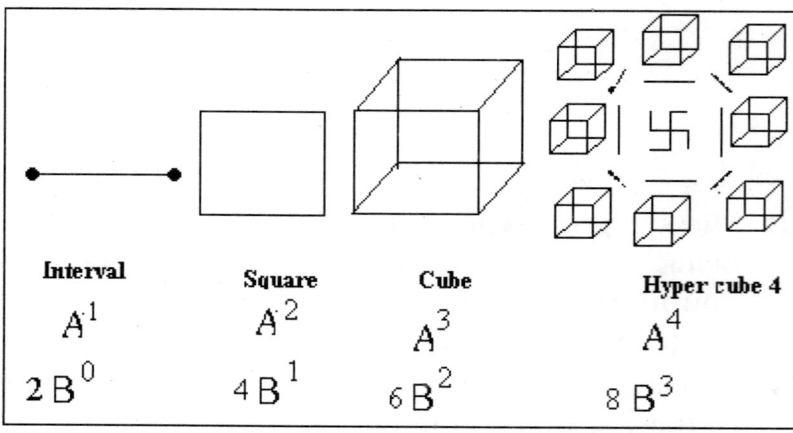

1. One shall have a fresh look at the set up of an interval and observe that, it has a pair of boundary points and a length in between. Length may be expressed as A^1 and the end / boundary points together may be expressed as 2 B^0.
2. A fresh look at the set up of the square shall be taking us to four boundary lines, which together

may be expressed as $4B^1$ and the area of surface in between the boundary lines may be expressed as A^2.

3. The working rule of Ganita Sutra-1, 'one more than before', shall be taking us from length / domain of an interval A^1 to area / domain of a square A^2.
4. Likewise the working rule of Ganita Upsutra-1 'to follow the proportionate symmetry of form within frame, shall be taking us from 2 boundary components of interval to 4 boundary components of square.
5. The point (B^0) as boundary component of interval and line (B^1) as boundary component of square are also obeying the working rule of Ganita Sutra-1.
6. 2 B^0 as total boundary of an interval, and 4 B^1 as total boundary of square are obeying the rules of both, Ganita Sutra-1 as well as of Ganita Upsutra-1.
7. As such one may see that the domain boundary ratio of interval A^1 : 2 B^0 takes us to domain boundary ratio of square A^2 : 4 B^1 and that the same follows the rules of Ganita Sutra-1 and Ganita Sutra-2.
8. On the lines, it would be a good exercise to reach from the domain boundary ratio of square A^2 : 4 B^1 to the domain boundary ratio of cube A^3 : 6 B^2 following the rules of Ganita Sutra-1 and Ganita Sutra-2.
9. It further would be a good exercise to reach from the domain boundary ratio of cube A^3 : 6 B^2 to the domain boundary ratio of hyper cube 4 A^4 : 8 B^3.

10. As such, it would be a very good exercise to reach at the general domain boundary formulation for all the four bodies namely interval, square, cube and hyper cube4 as $A^N : 2N\ B^{N-1}$, N=1, 2, 3, 4.

■

Special focus ascpect

'UNIT TO UNITY'

Creator's space (4-space / B) is a spatial order space. Being a spatial order space, it has 2-space in the role of dimension. With it a pair of axes are available at dimensional level itself.

The availability of a pair of axes at dimensional level itself, shall be equipping the coordination arrangments within such a dimensional order which are to be of distinct organisation format than that of the linear order set up of a 3-space with a cube as its representative regular body.

This distinction may well be chased in terms of the distinguishing features of the set ups of 'cube' and 'hyper cube-4' respectively being the representative regular bodies of 3-space and 4-space.

Viewing from 3-space, the most glaring difference between the set ups of 'cube' and 'hyper cube-4' would be coming wihtin comprehension range of 'sensory field', as that, the pre-domiant expression of 'cube A' is of its volume / solid / domain fold, while, in the context, that of hyper cube-4 B is to be as of 'origin fold'.

However, while viewing from '4-space', the distinguishing feature of a cube from that of hyper cube-4, the dominating expression would be as that the cube is reaching up till the boundary of hyper cube-4.

As such, when the set ups of cube and hyper cube-4 are to be taken up for their comparative chase, the same, as 'domain folds' shall be leading us to the dimensional order features, as much as that the cube, as a domain fold is a set up of linear order while the hyper cube-4, as a domain fold, is a set up of a spatial order.

With this, the macro state for cube as is its domain fold, the same is of a linear order. On the other hand, the macro state for hyper cube 4, as is its domain fold, the same is of a spatial order.

In the light of this predominant distinguishing features of a cube and of hyper cube 4, and that too at macro state level, it may be a blissful chase to follow these set ups, as such, in terms of their dimensional orders, which in case of a cube being of linear features and the same in case of hyper cube-4 being of spatial features, as such the whole focus, naturally, that way gets centred at the properties and characteristics which distinguish spatial order from that of linear order.

The spatial order, predominantly, with the privilege of having a pair of axes, the same, naturally has an advantage over the linear order which is to be satisfied with only a single axis. This advantage of a pair of axes over a single axis, amongst other features, is to be permitting its chase in terms of artifices as being of 'count' to for 'spatial order at coutn 1' for a linear order. Further, the same, as an operation, the spatial order is to be approached as 'PAIRING'.

The shift from artifice '1' to artifice '2', and consequential shift from 'count' to 'count as pairing', becomes the basic aspect of the ordering system which deserves to be glimpsed, comprehended and chased.

This, when viewed from 3-space, as a linear order, the same is to emerge as of the range from 'cube' to cube end as 'cuboid'. However, the same when is to be viewed from the spatial order point of view, it is to take us to the range of artifice 6, being of the features 3+3. The other features of artifice 6 being as that 6=3+3=2+2+2=1+2+3=1 x 2 x 3 which makes it a perfect number, the same are there becasue of the pairing operation of artifices.

One chase path for the pairing operation of artifices of numbers, is that, the count 1 as number 1 manifests for it a single artifice. And the same manifests a pair of artifices for 'count 2 / number 2'. These together sum up 'triple artifices'. This emergence, this way takes to the emerging range as being of a set of six artifices.

This feature of pairing operation, is having its full expression within Creator's space / 4-space and here within this domain there being a dimensional frame of quadruple spatial dimensions, as such there emerges a four fold manifestation format with each fold itself being of a spatial order, and thereby the sequential unfoldment from 'whole' to 'parts', being inherently there, and the same is to be taken as being there because of the 'pairing operation', which takes a pair of sequential steps at a time.

The potentialities of pairing operation are there in its feature of 'taking a pair of sequential steps at a time 'which makes the spatiai order ssytem taking care of itself by simultaneoulsy processing availing a pair of axes, of which while one of them accepts increasing '**doubling**' progression, the other on the other hand accept decreasing '**halving**' progression.

The simultaneous handling of 'doubling' as well as 'halving' permits the unit to maintain its internal unity intact, as that 2 x 1/2 =1 x 1=1. It is because of this potentiality of the spatial order (2-space in the role of dimension) of the Creator's space (4-space), that the static cube with availability of a degree of freedom of motion within hyper cube 4, acquires an additional edge, that is 13th edge, with which, cube as along a spatial format, shall be a set up of 'artifice 26', and further the spatial format being there because of the spatial order of creator's space, it shall be providing a manifestation format for the set up of the cube, and the same, accordingly, deserves to be chased as such.

3-SPACE MATHEMATICS, SCIENCES AND TECHNOLOGIES

1. '3-space' with focus upon its linear dimensions makes mathematics, sciences and technologies' of this space being of 'linear features'.
2. Cube as representative regular body of 3-space

with 'Volume of a cube' as dominant expression being domain fold of this as manifestation layer makes 'mathematics, sciences and technologies' of this space, being of 'physical content lump' features.

3. From minutest granule to the biggest solid, the whole range, being of same physical content features as well as of identical linear dimensional axes characteristics, as such within each point of 3-space stands inherently in-built a three dimensional frame so 'the mathematics, sciences and technologies' of this space are of 'macro state' features.
4. As the origin from which the axes of 3 dimensional frame remaining sealed and dormant at 0 value, so 'mathematics, sciences and technologies' of this space essentially remain to be of 'monad nature' of linear quantifiers.
5. The monad nature essentially being of features of 'monad' always coming into play 'as a whole', therefore 'the mathematics, sciences and technologies' of this space are to be of 'positive features', and conceptually, the sequential progression is to be of 'one directional flow' alone.
6. Being of one directional flow alone, as such, its processing potentialities are to be only of 'half range', and other half range is always to remain un covered by its processing systems.
7. The working out of second half of the range in terms of the first half of the range makes the processing of this space 'as of forced symmetries',

sacrificing the zero and negative artifices attributes.

8. The working bias of 'forced symmetry' deserves to be compensated failing which the Reality is bound to go guised and its restoration is bound to be accepted as an impossibility.

9. The Reality so processed under forced symmetric acceptance is to deprive of not only of the existence of four and higher dimensional spaces but also the lower, negative, zero as well as one and two spaces as well have to remain beyond reach.

10. The range of 'two and one' spaces would get disguised underneath the processing steps of first and second of the 3 axes of three-dimensional frame of 3-space, and the processing in respect of one and two space, as such would be of those spaces within 3-space, and not as of their independent existence as 1-space and 2-space respectively.

11. Euclids and Descartes were duped, and all of us, not being conscious of this trap, are also the easy preys of it.

12. The answer and remedy for ensured escape is to first concentrate upon the Creator's space (4-space), and then the cube as representative regular body of 3-space be approached as creation manifesting along the four folds manifestation format supplied by Creator's space.

13. 'The cube' as manifested creation is of four folds, of which 1-space is playing the role of dimension and its content lump is manifesting as the dimension fold.

14. '2-space' is playing the role of boundary and its content lump is manifesting as the boundary fold.
15. '3-space' is playing the role of domain and its content lump is manifesting as the domain fold.
16. '4-space' is playing the role of origin and its content lump is manifesting as the origin fold.
17. A step ahead, the chase is to be of the way the transcendental base of Creator's space fulfills the Creator's space with solid quantifiers, and the cube has its distinct role but as along the transcendence range of five folds, the fifth in sequence being the role of 5-space itself otherwise lively at the base of the Creator's space.
18. As such, the sadkhas fulfilled with an intensity of urge to fully know and completely chase the different roles of '3-space' and hence 'the mathematics, sciences and technologies of 3-space' shall permit their mind to transcend following the way Creator's space is fulfilled with transcendental features and the way creations automatically attain self referral features.

TO CHASE THE MANIFESTATION OF CUBE

1. Vedic mathematics, sciences and technologies of 3-space, essentially are to chase the manifestations of solids and in particular the cubes, spheres and cones, prisms and pyramids with rectangular and polygon formats, and â•˜this specific shapes and forms features of manifestations of 3-space bodiesâ•™ as such, as well may be designated as chase of CRYSTALS, DIAMONDS and GEMS, as Macro states.

2. Of these, the beginning shall be had with manifestation of CUBE.

3. The manifestation of cube as a set-up within geometric envelope stitched as of 8 corner points, 12 edges and 6 surfaces, in all, 26 geometric components is the first basic feature, which need be chased.

4. Each of the 8 corner points accepts the role of origin for a three-dimensional frame of half dimensions, and thereby, there is available, as many as 8 such frames, which together, shall be constituting a set of 4 three dimensional frames.

5. The coordination of 8 three dimensional frames of half dimensions as a set of 4 three dimensional frames of full dimensions is a feature which deserves to be chased with a focus as the cube as a representative regular body of 3-space shall be having only 3 dimensions and these three dimensions can supply at the most a set of six half dimensions, while the requirement of all the half dimensions for 8 corner points is of 8 x 3 =24 half dimensions.

6. The split for three dimensional frame, even as of three solid dimensions, shall be making available only a set of three pairs of three dimensional frames of half dimensions. It shall, as such be expecting one another pair of three-dimensional frames, and the same, in the circumstances may be hoped having been supplied by the â•˜spaceâ•™ itself. The supply of a pair of three-dimensional frames of half dimensions of solid order is to be there only from the transcendental worlds (5-space) of solid dimensional order.

7. The split of a three dimensional frame of solid dimensional order into a pair of such dimensional frames is to be there because of the availability of Creator's space at the origin.
8. It is this availability of the Creator's space at the origin which need be chased along its all the four spatial dimensions, and it is this chase within spatial dimensions shall be providing a format of a pair of orientations permitting inter change there of, because of which the split of a three dimensional frame into a pair of three dimensional frame shall also be permitting reversal as well as translation of them for their setting along the corner points of the cube.
9. It is this pairing of the corner points and simultaneously insertion of a pair of three dimensional frames of half dimensions obtained simultaneously with split of a three dimensional frame and also reversal translation from them to reach and to be established within so paired corner points of the cube, and thereby, making the set up of the cube ensured for its all the four pair of corner points as a real solid set up.
10. The above feature of stitching of the set up of a cube by pairing corner points and embedding three dimensional frames in all the 8 corner points further ensures the stitching for all the twelve edges of the cube as of di monad format with pair of parts of the edge of the cube to be supplied by two different three dimensional frames of corner points coordinated by the concerned edge.
11. This feature of pairing of 24 half dimensions into

twelve edges, further in a sequence pairs twelve edges into six spatial frames and thereby is ensured the required six surface plates for the geometric envelope of the cube.

12. This sequential pairing in a pair of steps from 24 half dimensions to 12 edges to 6 surfaces, in that sequence and order or reverse thereof taking from six surfaces to 12 edges to 24 half dimensions to 8 three dimensional frame of half dimensions to 4 three dimensional frame of full dimensions are the sequential progressions which need be chased for replicating the Vedic mathematics, sciences and technologies.

13. For this exercise, another feature which need be taken account of is the exhaustive coverage for all the 8 corner points in terms of only 7 of the edges, to be designated as the manifested edges, while the remaining five edges to remain un manifested support.

14. This sequential progression for coordination of 8 corner points in terms of 7 edges, is to be of 3 fold orientations along the 3 dimensions springing out from each corner points. The chase along any of such orientations along any of the dimensions at its end reach at the 8th corner in that sequence and order, naturally would also provide reversal for such progression from first corner to 8th corner into 8th corner to 1st corner. This way, two fold progressions, for both of orientations for each edge would make the synthetic stitching of the geometric envelope along the edges to be of manifestation formats of the order of (-1) space

playing the role of dimension for (+1) space, and as such the transition and transformation from macro states of edges to micro states of edges would be available. It is this availability and permissibility of transition and transformation from macro states to microstates for the edges would make '0-space' lively in the role of boundary fold for such manifestation. The 0-space in its role as of dimension fold for 2-space would inherently coordinate corner points as 0-space bodies with surfaces of the cube as 2-space bodies.

15. Such is the richness of the geometric envelope of the cube and replication of it, naturally, can be expected for supplying parallel richness for the mathematics, sciences and technologies of 3-space, provided the whole approach to the 3-space bodies is the way these avail their manifestation formats.

16. The simple progressions emerging from 2 corner points of an interval, 4 corner points of square and 8 corner points of cube, is just to sway away with it a forced symmetries by working up till half range intervals and covering second half parallel to it. The progression 2^N, for its values N= 0, 1, 2, 3, 4 as manifestation layer (0, 1, 2, 3) is to be of the values 1, 2, 4 and 8. There is a jump over artifices '3 and 5, 6, 7'. This jump, would be a jump along third, fifth, sixth and seventh edges coordination of 8 corner points, and parallel to it would be a jump in respect of third, fifth, sixth and seventh geometries of 3-space.

17. The chase of seven edges coordination of a cube,

would firstly reverse the orientations from that of first edge to that of third edge, and then further the circular orientation of first three edges would have reversal for it along the last three edges, namely, along fifth, sixth and seventh edges. It is this reversal of orientations firstly at third and secondly at fifth, sixth and seventh edges, which would go disguised for working only with half of interval. The forcing of symmetry by working with half interval is going to be at such heavy structural cost.

18. The seven geometries of 3-space, accept classification as 3 positive geometries, 3 negative geometries and one zero signature geometry. The neutrality of orientation for zero signature geometry is to cause slip with it being its own negative and there by there being four non positive and also four non negative geometries. The working with half interval, as such, in the context is bound to be at the cost of four out of seven geometries in all of three space.

Pairing to squaring

More insight into the pairing process at the base of manifestation format may be had by chasing the 'squaring' as a word formulation which accepts number value format for it as of artifice 106.

Conceptually, **number value format** (in short **NVF**) is the expression of pairing of artifices 1 to 26 which in the context of English alphabet amounts to alphabet letters A to Z accepting association of artifices 1 to 26 in that sequence and order, and then the words availing particular alphabet letters to yield combined number value format for the whole formulation of the word as summation of the individual number value formats of the concerned individual letters constituting the word formulations.

This, this way, worksout NVF (SQUARING) = 19 + 17 + 21 + 1 + 18 + 9 + 14 + 7 = 106. The artifice 106 accepts reorganistaion as 53+53. This pairing (53, 53) is number values formats pairing NVF (SQUARING) =NVF (AXIS, AXIS).

In the light of above insight, it may help appreciate as to how linear ordered set up, is to make 'NVF' (LINEAR) = NVF (SOLID); and NVF (SQUARE)= 81= 9 X 9= 3^2 x 3^2, a paired pairing which shall be taking us to 4-space being a spatial ordered space of 4 spatial dimensions; a creator's space of pairing operation constituting NVF (LINEAR) = NVF (SOLID) In the context, it also would be relevant to note as that, how blissful is the set up of the Creator's space where NVF (MONAD, MONAD)= NVF (TRI MONAD).

Availing rich resource of words formulations of pairing operation

The orthodox and classical words formulations is a big resource which deserves to be availed fully. At initial stage, these words formulations may be availed for having insight about the partitions of artifices at the base of NVFs of these words formulations.

The second way to avail these words formulations is to have insight about the conceptual terms like pairing, squaring and so so on in terms of their number value format.

Illustratively NVF (PAIRING) = 74 = NVF (CONE, CONE) and NVF (SQUARING)=106 = NVF (AXIS, AXIS). Further NVF (PAIRING)= 74 =NVF (ENGLISH(, which as such makes 'pairing' as the basic operation of these words formulations. The partitions of artififce 74 shall further be giving us insight of internal structural characteristics of the pairing features.

Illustratively 74=1+73 amounts to NVF (PAIRING) = NVF 9 (A FORMAT).

Like that it shall be taking us up till NVF (PAIRING) = NVF (CONE, CONE). Then further NVF (SERIES) = 75 = NVF (A PAIRING).

Then, 74 as one of the components of artifices 75 onwards shall be supplying us, the whole range of external features of PAIRING.

The chase like this shall be perfecting intelligence and the same deserves to be availed. The book Vedic mathematics decodes SPACE BOOK may give further insight into this aspect of Vedic mathematics.

It would be appropriate to impress for intensity for the urge to chase the pairing operation agailed by word formulation by taking up the illutration of NVF (PERFECTION) = 111 = NVF (CONE, CONE, CONE) = NVF (TETRA MONAD).

It would be blissful for the sadkhas to perfect their intelligence by chasing the pairing operation availed by the words formulations, begining as:

NVF (PAIRING) = NVF (CONE, CONE)

NVF (MONAD, MONAD) = NVF (TRI MONAD)

NVF (PERFECTION) = NVF (CONE, CONE, CONE)

= NVF (TETRA MONAD)

NVF (UNIT) = NVF (ZERO), a situation where 'zero' and unit are of 'same' format. One may evaluate and indexed ones perfection in terms of ones comprehension and appreciation of 0 and 1 being of same format and unity range.

■

7
OVERVIEW

Vedic mathematics is the ancient wisdom discipline. The appropriate age range for capture of its basics is 'between 10 to 14 years of age'.

Within the present day schedule of schooling, this crucial age group of '10 to 14 years of age' is the middle school teaching phase. As such the present book-1 to 5 series, as chase of Vedic mathematics basics, is to cover phase and stage of learning at middle school schedule.

The first aspect of any Discipline of instructions is 'the art of imparting instructions, that is, teaching'. It is only in terms of this principle fold of the Discipline, as that all other folds of the Discipline are to come within the comprehensions and applications thereof at the level of young minds, the learners.

It is the art of teaching only which really brings the science of learning within the comprehension and applications range of young minds.

Because of it, Book-1 is devoted to the art of teaching of Vedic mathematics, while Book-2 stands devoted to the 'science of learning' of Vedic mathematics.

Both art of Teaching and Science of learning are utmost important for the learners acquiring the values of the Discipline. The Art of Teaching is 'teachers centred', while 'the Science of Learning' is students centred.S

With this, the phase and stage of values of the Discipline takes to the subject content of the Discipline, whose 'teaching and learning' has been the concern of first two steps / folds of the Discipline. The subject content aspect, as third step in the series, focuses upon the 'basics of knowledge and wisdom' to be learnt and imbibed.

The pure and applied values may be deserved to be chased as distinct streams, of it, the first as 'basis of basics' and second as 'skills, to be chased on first principles'.

As such, book-3 of the series is devoted to 'basics basis' of Vedic mathematics and book-4 is devoted to 'The Skills of Vedic mathematics' for Perfection of Intelligence.

Finally, reaches a stage of glimpsing the 'Reality' which transcends our Triloki / Jagat / Vishwa / 3-space / Set up of a cube / linear order and is to take us to spatial order / hyper cube4 / 4-space / Creator's space / Brahma's domain.

OVERVIEW

The stepwise overview of this chase, beginning with Book-1 titled 'The Teaching of Vedic mathematics', which is to be teacher's centred step, takes us to the 'the foundation of the Discipline' as that the solid quantifiers, which ultimately manifest as our Triloki /

Jagat / Vishwa / 3-space / Set up of a cube / linear order, emerge within the Creator's space through its transcendental core.

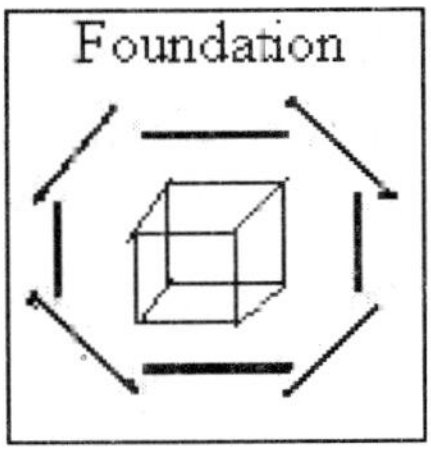

'The Teachers of Vedic mathematics' are to be at this state of chase of emergence of solid quantifiers within Creator's space through its transcendental core and their manifestation as our Triloki / Jagat / Vishwa / 3-space / Set up of a cube / linear order.

As such, the overview and sum up of book-1, The Art of Teaching Vedic mathematics, is to be chased in terms of the set up of 'foundation' as the transcendental worlds manifesting within creator's space, as the 'beginning as well as the end of this art'.

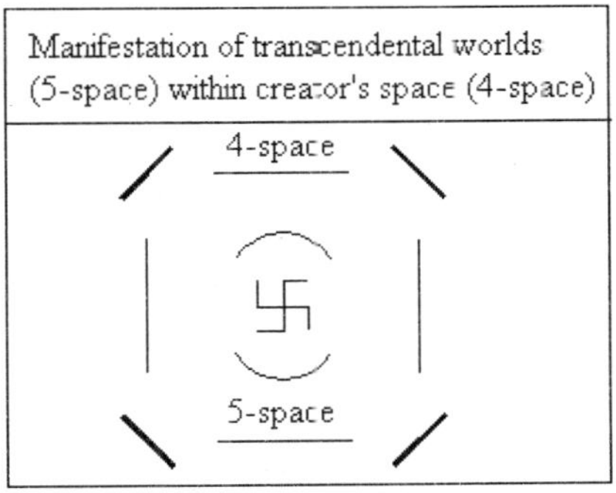

The science of learning, as the second step of the chase of Vedic mathematics, may have its starting point, focusing upon 'cube', beginning with its static state as a placement of

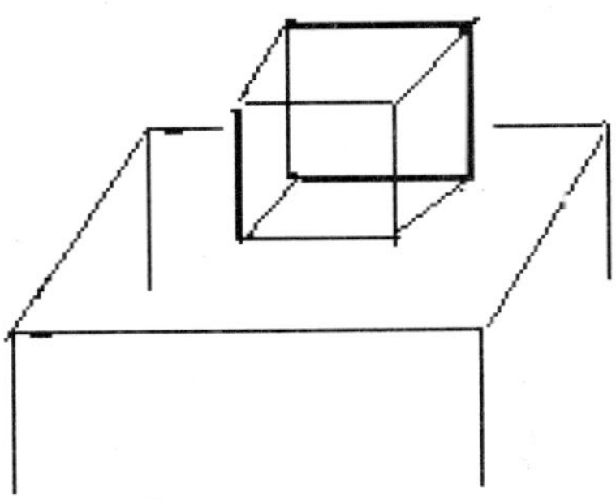

a cube upon a four legged table.

The teachers mental state, as such is taken as of wisdom as one attains while taking existence domain for oneself as of hyper cube4.

The students mental state, as comparison to it, is to be taken as of cube as the existence domain. The aim of teaching and learning is to be to attain a transition for the start with mental state of linear order of 3-space set up as cube to that of spatial order of 4-space set up as hyper cube4.

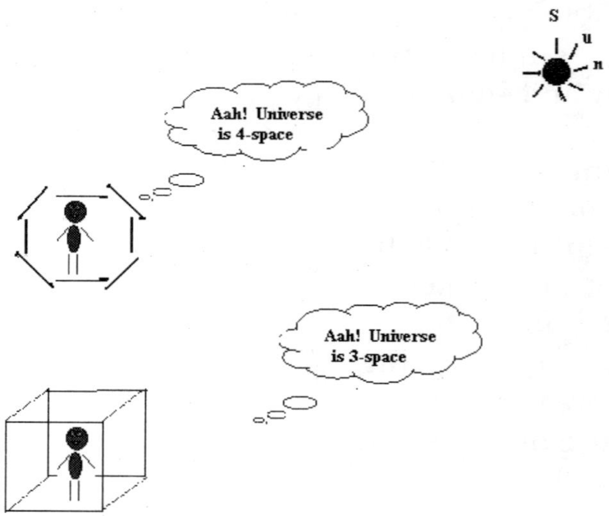

The subject content is to attain transition from the set up of a cube to the set up of hyper cube4. As such, step one for its attainment is to be to reach at the common basis of the basics of all branches of 'first, first two and all the three axes of the set up of cube / 3-space, and also of geometry, algebra and arithmetic', and there after, as a second step, the skills of chase along the coordination arrangement of different organization formats of set up of cube / 3-space for different operations of artifices of numbers, flow streams for the content within 3-space domain and the manifestations of 3-space content lump itself, are to be acquired.

This pair of steps, in the background of 'the art of teaching and science of learning steps, would emerge to be the 'third and fourth folds respectively of 'Vedic mathematics chase' being the subject content of book-3 and book-4 of this series.

The starting point of book-3 chase step is to be the common formulation of the geometric entities of interval, square and cube being $(X^1+2X^0)^N$, N=1, 2, 3.

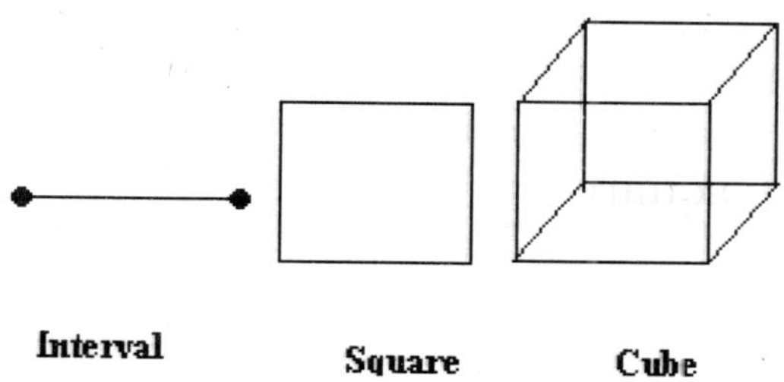

$$(X^1 + 2X^0)^N, N=1, 2, 3$$

$$(X^1 + 2X^0)^1 = X^1 + 2X^0$$

$$(X^1 + 2X^0)^2 = X^2+4X^1 + 4X^0$$

$$(X^1 + 2X^0)^3 = X^3 +6X^2+12X^1 + 8X^0$$

In a series, the phase and stage of book-4 for acquisition of skills for the knowledge of the coordination arrangement of organization formats of set up of a cube, shall be taking us to begin with the grid frames for the surfaces of the cube, and to have super imposition of grids to chase flow from the centres of the grid zones attaining solid domain of the cube.

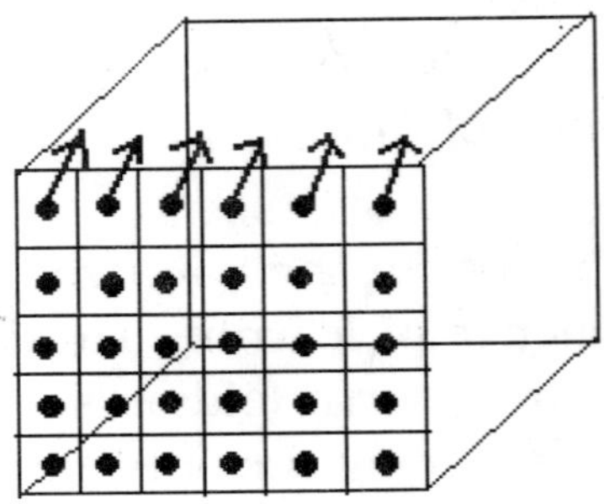

Finally, a stage is reached to glimpse the Reality transcending the set up of a cube / 3-space and the *sadhka* feels the melting of the mental frame of 3-space and the Reality of 4-space existence domain coming within the comprehension range. The starting point of this glimpse is to be hyper cube4 as well accepting common formulation along with the set ups of interval, square and cube:

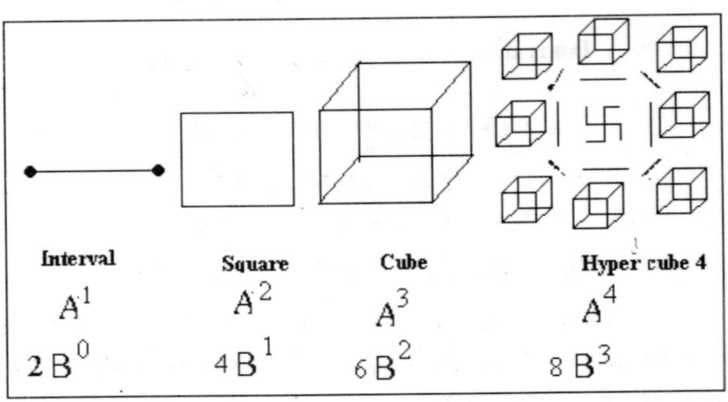

■